NEW YORK & PENNSYLVANIA

Tamara Eder

with contributions from Ian Sheldon

© 2001 by Lone Pine Publishing
First printed in 2001 10 9 8 7 6 5 4 3 2 1
Printed in Canada

THE PUBLISHER: LONE PINE PUBLISHING

1901 Raymond Avenue SW, Suite C	10145-81 Avenue
Renton, WA 98055	Edmonton, AB T6E 1W9
USA	Canada

Website: http://www.lonepinepublishing.com

National Library of Canada Cataloguing in Publication Data

Eder, Tamara, (date)
  Animal tracks of New York and Pennsylvania

  ISBN 1-55105-311-X

  1. Animal tracks—New York (State)—Identification. 2. Animal tracks—
Pennsylvania—Identification. I. Sheldon, Ian, (date) II. Title.
QL768.E365 2001   591.47'9   C2001-910026-4

Editorial Director: Nancy Foulds
Editor: Volker Bodegom
Proofreaders: Randy Williams, Lee Craig
Production Manager: Jody Reekie
Design, layout and production: Volker Bodegom, Monica Triska
Cover Design: Robert Weidemann
Cartography: Volker Bodegom
Animal illustrations: by Gary Ross, except for those by Ian Sheldon
    (p. 137), Kindrie Grove (p. 101), Ted Nordhagen (p. 127) and
    Ewa Pluciennik (p. 123).
Track illustrations: Ian Sheldon
Cover illustration: Raccoon by Gary Ross

We acknowledge the financial support of the Government of Canada
through the Book Publishing Industry Development Program (BPIDP)
for our publishing activities.

PC: P4

CONTENTS

INTRODUCTION

If you have ever spent time with an experienced tracker, or perhaps a veteran hunter, then you know just how much there is to learn about the subject of tracking and just how exciting the challenge of tracking animals can be. Maybe you think that tracking is no fun, because all you get to see is the animal's prints. What about the animal itself—is that not much more exciting? Well, for most of us who don't spend a great deal of time in the beautiful wilderness of New York and Pennsylvania, the chances of seeing the great Black Bear or the fun-loving River Otter are slim. The closest that we may ever get to some animals will be through their tracks, and they can inspire a very intimate experience. Remember, you are following in the footsteps of the unseen—animals that are in pursuit of prey, or perhaps being pursued as prey.

This book offers an introduction to the complex world of tracking animals. Sometimes tracking is easy. At other times it is an incredible challenge that leaves you wondering just what animal made those unusual tracks. Take this book into the field with you, and it can provide some help with the first steps to identification. Animal tracks and trails are this book's focus; you will learn to recognize subtle differences for both. There are, of course, many additional signs to consider, such as scat and food caches, all of which help you to understand the animal that you are tracking.

It takes many years to become an expert tracker. Tracking is one of those skills that grows with you as you acquire new knowledge in new situations. Most importantly, you will have an intimate experience with nature. You will learn the secrets of the seldom seen. The more you discover, the more you will want to know. And, by developing a good understanding of tracking, you will gain an excellent appreciation of the intricacies and delights of the marvelous natural world.

How to Use This Book

Most importantly, take this book into the field with you! Relying on your memory is not an adequate way to identify tracks. Track identification has to be done in the field or with

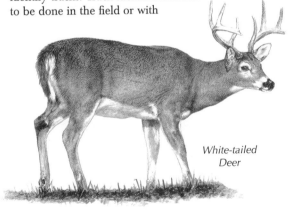

White-tailed Deer

detailed sketches and notes that you can take home. Much of the process of identification involves circumstantial evidence, so you will have much more success when standing beside the track.

This book is laid out in an easy-to-use format. Beginning on p. 140, there is a quick reference appendix to the tracks of all the animals illustrated in the book. The appendix is a fast way to familiarize yourself with certain tracks, and it guides you to the more informative descriptions of each animal and its tracks.

Each description is illustrated with the appropriate footprints and the track patterns that it usually leaves. Although these illustrations are not exhaustive, they do show the tracks or groups of prints that you will most likely see. You will find a list of dimensions for the tracks, giving the general range, but there will always be extremes, just as there are with people who have unusually small or large feet. Under the category 'Size' (of animal), the 'greater-than' sign (>) is used when the size difference between the sexes is pronounced.

If you think that you may have identified a track, check the 'Similar Species' section. This section is designed to help you confirm your conclusions by pointing out other animals that leave similar tracks and showing you ways to distinguish among them.

As you read this book, you will notice an abundance of words such as 'often,' 'mostly' and 'usually.' Unfortunately, tracking will never be an exact science; we cannot expect animals to conform to our expectations, so be prepared for the unpredictable.

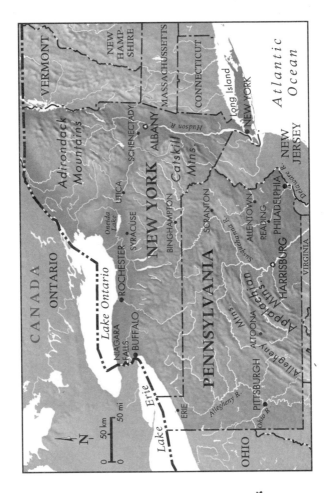

7

Tips on Tracking

As you flip through this guide, you will notice clear, well-formed prints. Do not be deceived! It is a rare track that will ever show so clearly. For a good, clear print, the perfect conditions are slightly wet, shallow snow that isn't melting or slightly soft mud that isn't actually wet. These conditions can be rare—most often you will be dealing with incomplete or faint prints, where you cannot even really be sure of the number of toes.

Should you find yourself looking at a clear print, then the job of identification is much easier. There are a number of key features to look for: measure the length and width of the print, count the number of toes, check for claw marks and note how far away they are from the body of the print, and look for a heel mark. Keep in mind more subtle features, such as the spacing between the toes, whether or not they are parallel and whether fur on the sole of the foot has made the print less clear.

When you are faced with the challenge of identifying an unclear print—or even if you think that you have made a successful identification from one print alone—look beyond the single footprint and search out others. Do not rely on the dimensions of just one print, but collect measurements from several prints to get an average impression. Even the prints within one trail can show a lot of variation.

Try to determine which is the fore print and which is the hind, and remember that many animals are built very differently from humans, having larger forefeet than hind feet. Sometimes the prints will overlap, or they can be directly on top of one another in a direct register. For some animals, the fore and hind prints are pretty much the same.

Check out the pattern that the tracks make together in the trail and follow the trail for as many paces as is necessary for you to become familiar with the pattern. Patterns are very important and can be the distinguishing feature between different animals with otherwise similar tracks.

Follow the trail for some distance—it can give you some vital clues. For example, the trail may lead you to a tree, indicating that the animal is a climber, or it may lead down into a burrow. This part of tracking can be the most rewarding, because you are following the life of the animal as it hunts, runs, walks, jumps, feeds or tries to escape a predator.

Take into consideration the habitat. Sometimes habitat alone will allow you to distinguish very similar tracks— one species might be found on riverbanks, whereas another might be encountered just in dense forest.

Think about your geographical location, too, because some animals have a limited range. This consideration can rule out some species and help you with your identification.

Remember that every animal will at some point leave a print or trail that looks just like the print or trail of a completely different animal!

Finally, keep in mind that if you track quietly, you might catch up with the maker of the prints.

Terms & Measurements

Some of the terms used in tracking can be rather confusing, and they often depend on personal interpretation. For example, what comes to your mind if you see the word 'hopping'? Perhaps you see a person hopping about on one leg, or perhaps you see a rabbit hopping through the countryside. Clearly, one person's perception of motion can be very different from another's. Some useful terms are explained on the next few pages, to clarify what is meant in this book, and, where appropriate, how the measurements given fit in with each term.

The following terms are sometimes used loosely and interchangeably—for example, a rabbit might be described as 'a hopper' and a squirrel as 'a bounder,' yet both leave the same pattern of prints in the same sequence.

Ambling: Fast, rolling walking.

Bounding: A gait of four-legged animals in which the two hind feet land simultaneously, usually registering in front of the fore prints. It is common in rodents and the rabbit family. 'Hopping' or 'jumping' can often be substituted.

Gait: Describes how an animal is moving at some point in time. Different gaits result in different observable trail characteristics.

Galloping: A gait used by animals with four legs of even length, such as dogs, moving at high speed, hind feet registering in front of forefeet.

Hopping: Similar to bounding. With four-legged animals, it is usually indicated by tight clusters of prints, fore prints set between and behind the hind prints. A bird hopping on two feet creates a series of paired tracks along its trail.

Loping: Like galloping but slower, with each foot falling independently and leaving a trail pattern that consists of groups of tracks in the sequence fore-hind-fore-hind, usually roughly in a line.

Mustelids (weasel family) often use **2×2 loping**, in which the hind feet register directly on the fore prints. The resulting pattern has angled, paired tracks.

Running: Like galloping, but applied generally to animals moving at high speed. Also used for two-legged animals.

Trotting: Faster than walking, slower than running. The diagonally opposite limbs move simultaneously; that is, the right forefoot with the left hind, then the left forefoot with the right hind. This gait is the natural one for canids (dog family), short-tailed shrews and voles.

Canids may use **side-trotting**, a fast trotting in which the hind end of the animal shifts to one side. The resulting track pattern has paired tracks, with all the fore prints on one side and all the hind prints on the other.

hind print

fore print

Walking: A slow gait in which each foot moves independently of the others, resulting in an alternating track pattern. This gait is common for felines (cat family) and deer, as well as wide-bodied animals, such as bears and porcupines. The term is also used for two-legged animals.

Other Tracking Terms:

Dewclaws: Two small, toe-like structures set above and behind the main foot of most hoofed animals.

Direct Register: The hind foot falls directly on the fore print.

double register   *direct register*

Double Register: The hind foot registers and overlaps the fore print only slightly or falls beside it, so that both prints can be seen at least in part.

Dragline: A line left in snow or mud by a foot or the tail dragging over the surface.

dragline

Gallop Group: A track pattern of four prints made at a gallop, usually with the hind feet registering in front of the forefeet (see '*galloping*' for illustration).

Height: Taken at the animal's shoulder.

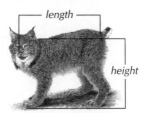

Length: The animal's body length from head to rump, not including the tail, unless otherwise indicated.

Metacarpal Pad: A small pad near the palm pad or between the palm pad and heel on the forefeet of bears and members of the weasel family.

Print (also called '***track***'): Fore and hind prints are treated individually. Print dimensions given are 'length' (including claws—maximum values may represent occasional heel register for some animals) and 'width.' A group of prints made by each of the animal's feet makes up a track pattern.

Register: To leave a mark—said about a foot, claw or other part of an animal's body.

Retractable: Describes claws that can be pulled in to keep them sharp, as with the cat family; these claws do not register in the prints. Foxes have semi-retractable claws.

Sitzmark: The mark left on the ground by an animal falling or jumping from a tree.

Straddle: The total width of the trail, all prints considered.

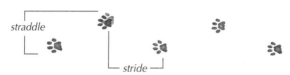

Stride: For consistency among different animals, the stride is taken as the distance from the center of one print (or print group) to the center of the next one. Some books may use the term 'pace.'

Track: Same as '**print**.'

Track Pattern: The pattern left after each foot registers once; a set of prints, such as a gallop group.

Trail: A series of track patterns; think of it as the path of the animal.

Porcupine

MAMMALS

River Otter

Moose

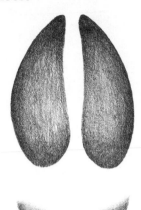

Fore and Hind Prints
Length: 4–7 in (10–18 cm)
Length with dewclaws: to 11 in (28 cm)
Width: 3.5–6 in (9–15 cm)

Straddle
8.5–20 in (22–50 cm)

Stride
Walking: 1.5–3 ft (45–90 cm)
Trotting: to 4 ft (1.2 m)

Size (bull>cow)
Height: 5–6.5 ft (1.5–2 m)
Length: 7–8.5 ft (2.1–2.6 m)

Weight
600–1100 lb (270–500 kg)

walking

MOOSE
Alces alces

The impressive male Moose, the largest of the deer, has a massive rack of antlers. Unfortunately, the Moose is critically endangered in both New York and Pennsylvania, so your chances of seeing one are slim.

Though ungainly in shape, the Moose moves gracefully, leaving a neat alternating walking pattern. The hind feet direct or double register on the fore prints. Long legs allow for easy movement in snow. Dewclaws—which give extra support for the animal's great weight—register in prints more than 1.2 inches (3 cm) deep, but far behind the hoof. In summer, look for tracks in mud beside ponds and other wet areas, where Moose especially like to feed; they are excellent swimmers. In winter, Moose feed in willow flats and coniferous forests, leaving a distinct browseline (highline). Ripped stems and scraped bark, 6 feet (1.8 m) or more above the ground, are additional signs of Moose.

Similar Species: White-tailed Deer (p. 20) tracks, though smaller, may be mistaken for juvenile Moose tracks. Feral Pig (p. 22) tracks are much smaller.

White-tailed Deer

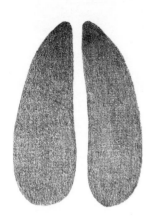

Fore and Hind Prints
Length: 2–3.5 in (5–9 cm)
Width: 1.6–2.5 in (4–6.5 cm)
Straddle
5–10 in (13–25 cm)
Stride
Walking: 10–20 in (25–50 cm)
Galloping: 6–15 ft (1.8–4.5 m)
Size (buck>doe)
Height: 3–3.5 ft (90–110 cm)
Length: to 6.3 ft (1.9 m)
Weight
120–350 lb (55–160 kg)

walking   *gallop group*

WHITE-TAILED DEER
Odocoileus virginianus

The keen hearing of this deer guarantees that it knows about you before you know about it. Frequently, all that we see is its conspicuous white tail in the distance as it gallops away, earning this deer the nickname 'Flagtail.' The adaptable White-tailed Deer can be found throughout New York and Pennsylvania, in small groups at the edges of forests and in brushlands. It can be common around ranches and residential areas.

This deer's prints are heart-shaped and pointed. Its alternating walking track pattern shows the hind prints direct registered or double registered on the fore prints. In snow, or when a deer gallops on soft surfaces, the dewclaws register. This flighty deer gallops in the usual style, leaving hind prints ahead of fore prints, with toes spread wide for steadier, safer footing.

Similar Species: Juvenile Moose (p. 18) tracks may be confused with large deer tracks. The Feral Pig (p. 22) makes similar tracks, but the tracks have a shorter stride and show a wider gap between the toes with dewclaws at the side instead of the rear.

Feral Pig

**Fore and Hind Prints
(with dewclaws)**
Length: 2.5–3 in (6.4–7.5 cm)
Width: 2.3 in (5.8 cm)
Straddle
5–6 in (13–15 cm)
Stride
Trotting: 16–20 in (40–50 cm)
Size (male>female)
Height: 3 ft (90 cm)
Length: 4.3–6 ft (1.3–1.8 m)
Weight
77–440 lb (35–200 kg)

trotting

FERAL PIG (Wild Pig, Wild Boar)

Sus scrofa

Descended from introduced European animals, the Feral Pig interbred with escaped Domestic Pigs (different in appearance but the same species). Populations of this sturdy beast of the dense undergrowth can be found in small, scattered populations in Pennsylvania. Armed with heavy tusks, they can be quite threatening.

A Feral Pig's print shows two prominent, widely spaced toe marks and usually (except perhaps on firm surfaces) distinct pointed dewclaw marks off to each side. The hind print is slightly smaller than the fore print. Feral Pigs are keen foragers, so their tracks can often be numerous, especially when they travel in a group. They usually trot—a typical track pattern shows a double register of hind print on fore print in an alternating pattern. Other signs are wallows and diggings.

Similar Species: White-tailed Deer (p. 20) tracks are slightly larger and pointier, with a narrower gap between the toes, a longer stride and dewclaw marks to the rear (not the side). Although the Domestic Pig is the same species, its tracks have a wider straddle and are less neat, often forming two separate lines.

Horse

Fore Print
(hind print is slightly smaller)
Length: 4.5–6 in (11–15 cm)
Width: 4.5–5.5 in (11–14 cm)
Straddle
2–7.5 in (5–19 cm)
Stride
Walking: 17–28 in (43–70 cm)
Size
Height: to 6 ft (1.8 m)
Weight
to 1500 lb (680 kg)

walking

HORSE
Equus caballus

Wilderness adventures on horseback are a popular activity, so you can expect horse tracks to show up almost anywhere.

Unlike any other animal in this book, the Horse has just one huge toe on each foot. This toe leaves an oval print with a distinctive 'frog' (V-shaped mark) at its base. If a Horse is shod, the horseshoe shows up clearly as a firm wall at the outside of the print. Not all horses are shod, so do not expect to see this outer wall on every horse print. A typical, unhurried horse trail shows an alternating walking pattern, with the hind prints registered on or behind the slightly larger fore prints. Horses are capable of a range of speeds—up to a full gallop—but most recreational horseback riders take a more leisurely outlook on life, preferring to walk their horses and soak up the beautiful views!

Similar Species: Mules (rarely shod) make smaller prints.

Black Bear

fore

hind

Fore Print
Length: 4–6.3 in (10–16 cm)
Width: 3.8–5.5 in (9.5–14 cm)

Hind Print
Length: 6–7 in (15–18 cm)
Width: 3.5–5.5 in (9–14 cm)

Straddle
9–15 in (23–38 cm)

Stride
Walking: 17–23 in (43–58 cm)

Size (male>female)
Height: 3–3.5 ft (90–110 cm)
Length: 5–6 ft (1.5–1.8 m)

Weight
200–600 lb (90–270 kg)

walking

BLACK BEAR
Ursus americanus

The Black Bear has a scattered range in forested areas throughout most of the region, but do not expect to see its tracks in winter, when it is dormant. Finding fresh bear tracks can be a thrill, but take care—the bear may be just ahead. Never underestimate the potential power of a surprised bear!

Black Bear prints somewhat resemble small human prints, but they are wider with claw marks. The small inner toe rarely registers. The forefoot's small heel pad often registers, and the hind print shows a big heel. The bear's slow walk results in a slightly pigeon-toed double register with the hind print on the fore print. More frequently, at a faster pace, the hind foot oversteps the forefoot. When a bear runs, the two hind feet register in front of the forefeet in an extended cluster. Along well-worn bear paths, look for 'digs' (patches of dug-up earth) and 'bear trees' whose scratched bark shows that this bear climbs.

Similar Species: No other animal in the region leaves clawed prints this big.

Domestic Dog

fore

hind

Fore Print
(hind print is smaller)
Length: 1–5.5 in (2.5–14 cm)
Width: 1–5 in (2.5–13 cm)
Straddle
1.5–8 in (3.8–20 cm)
Stride
Walking: 3–32 in (7.5–80 cm)
Loping to Galloping: to 9 ft (2.7 m)
Size
Very variable
Weight
Very variable

walking

loping to
galloping

DOMESTIC DOG
Canis familiaris

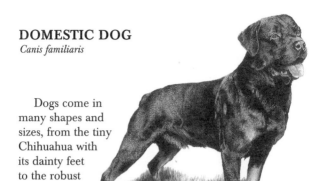

Dogs come in
many shapes and
sizes, from the tiny
Chihuahua with
its dainty feet
to the robust
and powerful
Great Dane.
Consequently, Domestic Dog tracks
vary enormously. Dog ownership is high
in many residential areas, and the popular pastime
of dog walking can result in many dog tracks being left
scattered about, especially if there is wet mud or snow.

The forefeet of the Domestic Dog, which are much
larger than the hind feet and support more of the animal's weight, leave the clearest tracks. When a dog
walks, the hind prints usually register ahead of or beside
the fore prints. As the dog moves faster, it trots and then
lopes before it gallops. In a trot or lope pattern the prints
alternate fore-hind-fore-hind, whereas a gallop group
shows (from back to front) fore-fore-hind-hind.

Similar Species: Keep in mind that dog prints are
usually found close to human tracks or activity. The
Coyote's (p. 30) more oval prints tend to splay less,
and its trail is more direct. Fox (pp. 32–35) tracks
may be confused with small dog tracks.

Coyote

fore

hind

Fore Print
(hind print is slightly smaller)
Length: 2.4–3.2 in (6–8 cm)
Width: 1.6–2.4 in (4–6 cm)

Straddle
4–7 in (10–18 cm)

Stride
Walking: 8–16 in (20–40 cm)
Trotting: 17–23 in (43–58 cm)
Galloping/Leaping:
    2.5–10 ft (0.8–3 m)

Size (female is slightly smaller)
Height: 23–26 in (58–65 cm)
Length: 32–40 in (80–100 cm)

Weight
20–50 lb (9–23 kg)

walking
or trotting

gallop
group

COYOTE
(Brush Wolf, Prairie Wolf)
Canis latrans

This widespread and adaptable canine prefers open grasslands or woodlands. On its own, with a mate or in a family pack, it hunts rodents and larger prey. A Coyote also occasionally develops an interesting cooperative relationship with a Badger (p. 60); you might find their tracks together where they have been digging for ground-dwelling rodents. If you find a Coyote den—usually a wide-mouthed tunnel leading into a nesting chamber—do not disturb it, or the female will move her pups.

The hind print is slightly smaller than the oval fore print, and its less triangular heel pad rarely registers clearly. Usually just the middle two toes register claw marks. The Coyote typically walks or trots in an alternating pattern; the walk has a wider straddle, and the trotting trail is often very straight. When it gallops, the Coyote's hind feet fall in front of its forefeet; the faster it goes, the straighter the gallop group. The Coyote's tail, which hangs down, leaves a dragline in deep snow.

Similar Species: A Domestic Dog's (p. 28) less oval prints splay more, and its trail is erratic. Fox (pp. 32–35) prints are usually smaller.

Red Fox

fore

hind

Fore Print
(hind print is slightly smaller)
Length: 2.1–3 in (5.3–7.5 cm)
Width: 1.6–2.3 in (4–5.8 cm)

Straddle
2–3.5 in (5–9 cm)

Stride
Trotting: 12–18 in (30–45 cm)
Side-trotting: 14–21 in (35–53 cm)

Size (vixen is slightly smaller)
Height: 14 in (35 cm)
Length: 22–25 in (55–65 cm)

Weight
7–15 lb (3.2–7 kg)

trotting | *side-trotting*

32

RED FOX
Vulpes vulpes

Very adaptable and intelligent, this beautiful and notoriously cunning fox is found throughout both states in a variety of habitats from forests to open areas.

Abundant foot hair allows just parts of the toes and heel pads to register, with no fine detail. The horizontal or slightly curved bar across the fore heel pad is diagnostic. A trotting Red Fox leaves a distinctive straight alternating trail—the hind print direct registers on the wider fore print. When the fox side-trots, its print pairs show the hind print to one side of the fore print in typical canid fashion. This fox gallops like the Coyote (p. 30). The faster the gallop, the straighter the gallop group.

Similar Species: Other canid prints lack the bar across the fore heel pad. Domestic Dog (p. 28) prints can be of similar size, but with a shorter stride and a less direct trail. Gray Fox (p. 34) prints are smaller. Small Coyote prints are similar, but they have a wider straddle, and the toe marks are more bulbous.

Gray Fox

fore

hind

Fore Print (hind print is slightly smaller)
Length: 1.3–2.1 in (3.3–5.3 cm)
Width: 1.1–1.5 in (2.8–3.8 cm)
Straddle
2–4 in (5–10 cm)
Stride
Walking/Trotting: 10–12 in (25–30 cm)
Size
Height: 14 in (35 cm)
Length: 21–30 in (53–75 cm)
Weight
7–15 lb (3.2–7 kg)

walking

GRAY FOX
Urocyon cinereoargenteus

 This small, shy fox can sometimes be seen in woodlands and chaparral country or in open areas that border forests. The Gray Fox is the only fox that climbs trees, which it does either for safety or to forage.

 The forefoot registers better than the smaller hind foot, and the hind foot's long, semi-retractable claws do not always register. The heel pads are often unclear—they sometimes show up just as small, round dots. When it walks, this fox leaves a neat alternating track pattern. When it trots, its prints fall in pairs, with the fore print set diagonally behind the hind print. The Gray Fox's gallop group is like the Coyote's (p. 30).

Similar Species: The Red Fox (p. 32) has barred fore heel pads; in general, its prints are larger and less clear (because of thick fur), and its stride is longer. Coyote and Domestic Dog (p. 28) prints are larger. Domestic Cat (p. 40) and Bobcat (p. 38) prints lack claw marks and have larger, less symmetrical heel pads.

Lynx

fore

hind

Fore Print (hind print is slightly smaller)
Length: 3.5–4.5 in (9–11 cm)
Width: 3.5–4.8 in (9–12 cm)

Straddle
6–9 in (15–23 cm)

Stride
Walking: 12–28 in (30–70 cm)

Size
Length: 2.5–3 ft (75–90 cm)

Weight
15–30 lb (7–14 kg)

walking

LYNX
Lynx canadensis

The elusive Lynx is found only in remote, dense forests, but its numbers have declined so rapidly in the last few decades that it is now critically endangered in this region. With huge feet and a relatively lightweight body, it stays on top of the snow as it pursues its main prey, the Snowshoe Hare (p. 66).

This cautious walker leaves a neat alternating track pattern, the hind print direct registered on top of the fore print. Thick fur on the feet often results in prints that are big, round depressions with no detail. In deeper snow, 'handles' off to the rear may extend the print. However deep the snow, this cat sinks no more than 8 inches (20 cm), and it rarely drags its feet. It is more likely to bound than to run. Its curious nature results in a meandering trail that may lead to a partially buried food cache.

Similar Species: Bobcat (p. 38) prints are smaller; its trail may show draglines. Fisher (p. 50) prints may not show the fifth toe—look for mustelid habits. Canid (pp. 28–35) prints show claw marks and their length exceeds their width.

Bobcat

fore

hind

**Fore Print
(hind print is slightly smaller)**
Length: 1.8–2.5 in (4.5–6.5 cm)
Width: 1.8–2.5 in (4.5–6.5 cm)

Straddle
4–7 in (10–18 cm)

Stride
Walking: 8–16 in (20–40 cm)
Running: 4–8 ft (1.2–2.4 m)

Size (female is slightly smaller)
Height: 20–22 in (50–55 cm)
Length: 25–30 in (65–75 cm)

Weight
15–35 lb (7–16 kg)

walking

*ambling
to loping*

BOBCAT
(Wildcat)
Lynx rufus

The Bobcat, a stealthy and usually nocturnal hunter, is seldom seen. Very adaptable, it can leave tracks anywhere from wild mountainsides to chaparral and even in residential areas.

A walking Bobcat's hind feet usually register directly on its larger fore prints. As the Bobcat picks up speed, its trail becomes an ambling pattern of paired prints, the hind leading the fore. At even greater speeds it leaves four-print groups in a lope pattern. The fore prints often exhibit asymmetry. The front part of the heel pad has two lobes and the rear part has three. In deep snow this cat's feet leave draglines. Half-buried scat along the Bobcat's meandering trail marks its territory.

Similar Species: Lynx (p. 36) prints can be similar. Large Domestic Cats (p. 40) have similar prints, but with a shorter stride and a narrower straddle. Canid (pp. 28–35) prints will be narrower than they are long and show claw marks, and the front of the footpad will have just one lobe.

Domestic Cat

fore

hind

Fore Print
(hind print is slightly smaller)
Length: 1–1.6 in (2.5–4 cm)
Width: 1–1.8 in (2.5–4.5 cm)

Straddle
2.4–4.5 in (6–11 cm)

Stride
Walking: 5–8 in (13–20 cm)
Loping/Galloping:
    14–32 in (35–80 cm)

Size (male>female)
Height: 20–22 in (50–55 cm)
Length with tail: 30 in (75 cm)

Weight
6.5–13 lb (3–6 kg)

walking

*loping to
galloping*

DOMESTIC CAT
(House Cat)
Felis catus

The tracks of the familiar and abundant Domestic Cat can show up almost any place where there are people. Abandoned cats may roam farther afield; these 'feral cats' lead a pretty wild and independent existence. Domestic Cats come in many shapes, sizes and colors.

As with all felines, a Domestic Cat's fore print and slightly smaller hind print both show four toe pads. Its retractable claws, kept clean and sharp for catching prey, do not register. Cat prints usually show a slight asymmetry, with one toe leading the others. A Domestic Cat makes a neat alternating walking track pattern, usually in direct register, as one would expect from this animal's fastidious nature. When a cat picks up speed, it leaves clusters of four prints, the hind feet registering in front of the forefeet.

Similar Species: A small Bobcat (p. 38) may leave tracks similar to a very large Domestic Cat's. Canid (pp. 28–35) prints show claw marks.

Raccoon

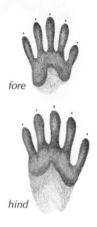

fore

hind

Fore Print
Length: 2–3 in (5–7.5 cm)
Width: 1.8–2.5 in (4.5–6.5 cm)

Hind Print
Length: 2.4–3.8 in (6–9.5 cm)
Width: 2–2.5 in (5–6.5 cm)

Straddle
3.3–6 in (8.5–15 cm)

Stride
Walking: 8–18 in (20–45 cm)
Bounding: 15–25 in (38–65 cm)

Size (female is slightly smaller)
Length: 24–37 in (60–95 cm)

Weight
11–35 lb (5–16 kg)

walking

*bounding
group*

RACCOON
Procyon lotor

The inquisitive Raccoon, common throughout these two states, is adored by some people for its distinctive face mask, yet disliked for its boundless curiosity—often demonstrated with residential garbage cans. A good place to look for its tracks is near water at low elevations. The Raccoon likes to rest in trees. It usually dens up for the colder months.

The Raccoon's unusual print, showing five well-formed toes, looks like a human handprint; its small claws appear as dots. Its highly dexterous forefeet rarely leave heel prints, but its hind prints (generally much clearer) do show heels. In this mammal's peculiar walking track pattern, the left fore print is next to the right hind print (or just in front) and vice versa. When it is in deep snow (which is rarely), the Raccoon may use a direct-registering walk. The Raccoon occasionally bounds, leaving clusters with the hind prints in front of the fore prints.

Similar Species: Raccoon prints are normally distinctive. In deep snow or mud, River Otter (p. 48), Fisher (p. 50) or Woodchuck (p. 78) tracks may look similar.

Opossum

fore

hind

Fore Print
Length: 2–2.3 in (5–5.8 cm)
Width: 2–2.3 in (5–5.8 cm)

Hind Print
Length: 2.5–3 in (6.5–7.5 cm)
Width: 2–3 in (5–7.5 cm)

Straddle
4–5 in (10–13 cm)

Stride
5–11 in (13–28 cm)

Size
Length: 2–2.5 ft (60–75 cm)

Weight
9–13 lb (4–6 kg)

walking   *fast walking*

OPOSSUM
Didelphis virginiana

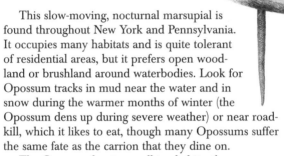

   This slow-moving, nocturnal marsupial is
found throughout New York and Pennsylvania.
It occupies many habitats and is quite tolerant
of residential areas, but it prefers open woodland or brushland around waterbodies. Look for
Opossum tracks in mud near the water and in
snow during the warmer months of winter (the
Opossum dens up during severe weather) or near roadkill, which it likes to eat, though many Opossums suffer
the same fate as the carrion that they dine on.
   The Opossum has two walking habits: the common
alternating pattern, with the hind prints registering on the
fore prints, and a Raccoon-like (p. 42) paired-print pattern, with each hind print next to the opposing fore print.
The very distinctive, long, inward-pointing thumb of the
hind foot does not make a claw mark. The Opossum is
an excellent climber. In snow, the long, naked tail's dragline may be bloodstained; this thinly haired animal is not
well adapted to the cold and frequently suffers frostbite.

Similar Species: Prints in which the distinctive thumbs
do not show may be mistaken for a Raccoon's.

Harbor Seal

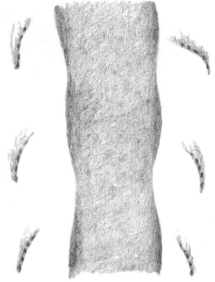

beach track

Size (male>female)
Length: 4–6 ft (1.2–1.8 m)
Weight
180–310 lb (80–140 kg)

HARBOR SEAL
Phoca vitulina

This seal, listed in New York State as vulnerable, can be found on coastal beaches. One of the smaller seals, it is quite shy and will usually slide off its rocky sentry post into the sea to escape curious naturalists. Look in sandy and muddy areas between platforms for its tracks—unmistakable because of their location and large size. Sometimes a Harbor Seal works its way up a river, so you may find its tracks along riverbanks.

Seals, though unrivaled in the water, are not the most graceful of animals on land. A seal's heavy, fat body and flipper-like feet can leave messy tracks: a wide trough, (made by its cumbersome belly) with dents alongside (made as the seal pushed itself along with its forefeet). Look for the marks made by the seal's nails.

Similar Species: There are no other seals in the region. The unusually shaped tracks of the Harbor Seal are unlikely to be confused with any other animal's.

River Otter

Fore Print
Length: 2.5–3.5 in (6.5–9 cm)
Width: 2–3 in (5–7.5 cm)

Hind Print
Length: 3–4 in (7.5–10 cm)
Width: 2.3–3.3 in (5.8–8.5 cm)

Straddle
4–9 in (10–23 cm)

Stride
Loping: 12–27 in (30–70 cm)

Size
(female is two-thirds the size of male)
Length with tail: 3–4.3 ft (90–130 cm)

Weight
10–25 lb (4.5–11 kg)

loping (fast)

RIVER OTTER
Lontra canadensis

No animal knows how to have more fun than a River Otter. If you are lucky enough to watch one at play, you will not soon forget the experience. Widespread and well-adapted for the aquatic environment, this otter lives near water; an otter in the forest is usually on its way to another waterbody. Expect to see a wealth of evidence of an otter's presence along the waterbodies in its home territory. The River Otter loves to slide in snow, often down riverbanks, leaving troughs nearly 1 foot (30 cm) wide; if there is no snow, it rolls and slides on grass and mud.

In soft mud, the webbing on the River Otter's five-toed feet, especially on the hind ones, may be evident. The inner toe of the hind foot is set slightly apart from the rest. If the forefoot's metacarpal pad registers, it lengthens the print. Very variable, otter trails usually show a typical mustelid 2×2 loping. However, with faster gaits they show three- and four-print groups. The thick, heavy tail often leaves a dragline.

Similar Species: Other mustelid trails do not show conspicuous tail drag. The Marten (p. 52) has similar-sized prints. Mink (p. 54) prints are about half the size. The Fisher (p. 50) usually has hairy feet, with forefeet larger than hind feet.

49

Fisher

Fore Print
Length: 2.1–4 in (5.3–10 cm)
Width: 2.1–3.3 in (5.3–8.5 cm)

Hind Print
Length: 2.1–3 in (5.3–7.5 cm)
Width: 2–3 in (5–7.5 cm)

Straddle
3–7 in (7.5–18 cm)

Stride
Walking: 7–14 in (18–35 cm)
Loping: 1–4.3 ft (30–130 cm)
Loping: 1–3 ft (30–90 cm)

Size (male>female)
Length with tail:
    34–40 in (85–100 cm)

Weight
3–12 lb (1.4–5.5 kg)

walking   *2×2 loping*

FISHER (Black Cat)
Martes pennanti

This agile hunter has declined dramatically in numbers to the point of being critically endangered in this region. The Fisher's speed and eager hunting antics make for exciting tracking as it races up trees and along the ground. It is one of the few predators capable of killing and eating Porcupines (p. 70).

All five toes may register, but the small inner toe frequently does not. Only the forefoot has a small heel pad that can show up in the print. The Fisher occasionally walks, making a direct-registering alternating track pattern, but it more often 2×2 lopes in typical mustelid fashion, leaving angled print pairs with the hind print direct registered on the fore print. Loping, its most common gait, produces three- and four- print groups (see the River Otter, p. 48). The patterns often vary within a short distance. The Fisher has been misnamed—it doesn't have much to do with water.

Similar Species: Male Marten (p. 52) tracks may be confused with a small female Fisher's, but Martens weigh less and leave shallower prints. Otters have larger hind feet than forefeet. When only four toes register, Fisher prints may look like a Bobcat's (p. 38).

Marten

Fore and Hind Prints
Length: 1.8–2.5 in (4.5–6.5 cm)
Width: 1.5–2.8 in (3.8–7 cm)

Straddle
2.5–4 in (6.5–10 cm)

Stride
Walking: 4–9 in (10–23 cm)
Loping: 9–46 in (23–120 cm)

Size (male>female)
Length with tail:
    21–28 in (53–70 cm)

Weight
1.5–2.8 lb (0.7–1.3 kg)

walking   *2×2 loping*

MARTEN
(American Sable)
Martes americana

This aggressive predator is found in coniferous and mixed-wood forests in parts of New York State. The Marten seldom leaves a clear print: often just four toes register, and the heel pad is undeveloped. The hairiness of the feet in winter often blurs pad detail, especially from the poorly developed palm pads. In the Marten's alternating walking pattern, the hind foot registers on the fore print. In 2×2 loping, the hind prints fall on the fore prints to form slightly angled print pairs in a typical mustelid pattern. The Marten's loping track patterns may appear as three- or four-print clusters (see the River Otter, p. 48). Follow the criss-crossing trails—if a Marten has scrambled up a tree, look for a sitzmark where it has jumped down.

Similar Species: Size and habitat are often key to distinguishing Marten, Fisher (p. 50) and Mink (p. 54) tracks. Female Fisher prints may resemble a large male Marten's but will be clearer. Male Mink prints overlap in size with small female Marten prints, but the Mink rarely climbs trees and (unlike the Marten) is usually found near water.

Mink

fore

hind

Fore and Hind Prints
Length: 1.3–2 in (3.3–5 cm)
Width: 1.3–1.8 in (3.3–4.5 cm)
Straddle
2.1–3.5 in (5.3–9 cm)
Stride
Walking/Loping: 8–36 in (20–90 cm)
Size (male>female)
Length with tail: 19–28 in (48–70 cm)
Weight
1.5–3.5 lb (0.7–1.6 kg)

2×2 loping

MINK
Mustela vison

The lustrous Mink is widespread but uncommon throughout these states. It prefers watery habitats surrounded by brush or forest. At home as much on land as in water, this nocturnal hunter can be exciting to track. Like the River Otter (p. 48), the Mink sometimes slides in snow, carving out troughs up to 6 inches (15 cm) wide.

The Mink's fore print shows five (perhaps four) toes, with five loosely connected palm pads in an arc, but the hind print shows only four palm pads. The metacarpal pad of the forefoot rarely registers, but the furred heel of the hind foot may register, lengthening the hind print. The Mink prefers the typical mustelid 2×2 loping gait, making consistently spaced, slightly angled double prints. Its diverse track patterns also include alternating walking; loping with three- and four-print groups (like the River Otter); and bounding (like a rabbit or hare, pp. 66–71).

Similar Species: Small Martens (p. 52) may have similar prints, but without a consistent 2×2 loping gait; they do not live near water. Weasel (pp. 56–59) tracks are similar but smaller. Bobcat (p. 38) prints may resemble four-toed Mink prints, but without claw marks or lobed palm pads, and they will not be in a bounding pattern.

Weasels

all weasels

Long-tailed Weasel
Fore and Hind Prints
Length: 1.1–1.8 in (2.8–4.5 cm)
Width: 0.8–1 in (2–2.5 cm)

Straddle
1.8–2.8 in (4.5–7 cm)

Stride
2×2 loping: 9.5–43 in (24–110 cm)

Size (male>female)
Length with tail:
    12–22 in (30–55 cm)

Weight
3–12 oz (85–340 g)

2×2 loping

LONG-TAILED WEASEL
Mustela frenata

Weasels are active,
year-round hunters with an avid
appetite for rodents. The Long-tailed
Weasel is the largest of the three weasels in the region.

Following a weasel's tracks can reveal much about the
activities of the nimble creature. Tracks are most evident
in winter, when weasels frequently burrow into the snow
or pursue rodents into their holes. Some weasel trails
may lead you up a tree. Weasels sometimes take to water.
To identify the weasel species, pay close attention to the
straddle, stride and loping patterns, and note the distri-
bution and habitat. The usual weasel gait is a 2×2 lope,
leaving a trail of paired prints. A weasel's light weight
and small, hairy feet result in pad detail that is often
unclear, especially in snow. Even with clear tracks, the
inner (fifth) toe rarely registers.

The Long-tailed Weasel's typical 2×2 lope shows an
irregular stride—sometimes short and sometimes long—
with no consistent behavior. Like the Mink (p. 54), this
weasel may bound like a rabbit or hare (pp. 66–71).

Similar Species: A large male Short-tailed Weasel's
(p. 58) tracks may be the same size as a small female
Long-tailed Weasel's. The Least Weasel (p. 58), with
smaller tracks, occupies some of the same habitats.

Weasels

Short-tailed Weasel
Fore and Hind Prints
Length: 0.8–1.3 in (2–3.3 cm)
Width: 0.5–0.6 in (1.3–1.5 cm)
Straddle
1–2.1 in (2.5–5.3 cm)
Stride
Loping: 9–36 in (23–90 cm)
Size (male>female)
Length with tail:
    8–14 in (20–35 cm)
Weight
1–6 oz (30–170 g)

Least Weasel
Fore and Hind Prints
Length: 0.5–0.8 in (1.3–2 cm)
Width: 0.4–0.5 in (1–1.3 cm)
Straddle
0.8–1.5 in (2–3.8 cm)
Stride
Loping: 5–20 in (13–50 cm)
Size (male>female)
Length with tail:
    6.5–9 in (17–23 cm)
Weight
1.3–2.3 oz (37–65 g)

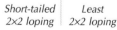

Short-tailed
2×2 loping

Least
2×2 loping

58

SHORT-TAILED WEASEL
(Ermine, Stoat)
Mustela erminea

The Short-tailed Weasel is smaller than the Long-tailed Weasel (p. 56) but larger than the Least Weasel (below). It prefers woodlands and meadows up to higher elevations but does not favor wetlands or dense coniferous forests.

This weasel's 2×2 loping tracks may fall in clusters, with alternating short and long strides.

Similar Species: A small female Long-tailed's (p. 56) tracks may be the same size as a large male Short-tailed's.

LEAST WEASEL
Mustela nivalis

The Least Weasel, found in small numbers in Pennsylvania, is the smallest weasel, with the least-clear tracks.

Its tracks may be found around wetlands and in open woodlands and fields.

Similar Species: A small female Short-tailed Weasel's (above) tracks may resemble a large male Least Weasel's, but the Short-tailed's habitats differ.

59

Badger

fore

hind

Fore Print (hind print is slightly shorter)
Length: 2.5–3 in (6.5–7.5 cm)
Width: 2.3–2.8 in (5.8–7 cm)

Straddle
4–7 in (10–18 cm)

Stride
Walking: 6–12 in (15–30 cm)

Size
Length: 21–36 in (53–90 cm)

Weight
13–25 lb (6–11 kg)

walking

BADGER
Taxidea taxus

This bold animal, with its squat shape and unmistakable face, is most often seen in the open grasslands, but the Badger also ventures into higher country. Thick shoulders and forelegs, coupled with long claws, make it a powerful digger. Look for Badger tracks in southern Pennsylvania, close to the Ohio border, especially in spring and fall snow—unlike most other mustelids, the Badger likes to den up in a hole for the coldest months of winter.

When a Badger walks, the alternating track pattern shows a double register, with the hind print sometimes falling just behind (or sometimes slightly in front of) the fore print. All five toes on each foot register. A Badger's long claws are evident in the pigeon-toed tracks that it leaves as it waddles along; the forefoot claws are longer than the hind-foot ones. In deep snow, the plowing action of the Badger's wide, low body often wipes out track detail.

Similar Species: In snow, a Porcupine (p. 72) trail may be similar, but it will show draglines made by the tail and quills, and it will likely lead up a tree, not to a hole.

Striped Skunk

fore

hind

Fore Print
Length: 1.5–2.2 in (3.8–5.5 cm)
Width: 1–1.5 in (2.5–3.8 cm)

Hind Print
Length: 1.5–2.5 in (3.8–6.5 cm)
Width: 1–1.5 in (2.5–3.8 cm)

Straddle
2.8–4.5 in (7–11 cm)

Stride
Walking/Bounding:
    2.5–8 in (6.5–20 cm)

Size
Length with tail:
    20–32 in (50–80 cm)

Weight
6–14 lb (2.7–6.5 kg)

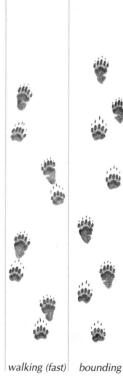

walking (fast)   *bounding*

STRIPED SKUNK
Mephitis mephitis

This striking skunk is notorious for its vile smell, and the lingering odor is often the best sign of its presence. Widespread throughout these two states in a diversity of habitats, it prefers lower elevations. The Striped Skunk dens up in winter, coming out on warmer days and in spring.

Forefeet and hind feet each have five toes. The long claws on the forefeet often register. The smooth palm pads and small heel pads leave surprisingly small prints. A skunk mostly walks—with such a potent smell for its defense, it rarely needs to run. Note that this skunk's trail rarely shows any consistent pattern, but an alternating walking pattern may be evident. The greater a skunk's speed, the more the hind foot oversteps the fore. If it runs, its trail consists of clumsy, closely set four-print groups. In snow it drags its feet.

Similar Species: The Eastern Spotted Skunk (p. 64) makes smaller prints in a very random pattern. Similar-sized mustelid (pp. 50–61) tracks will be farther apart than those resulting from a skunk's shuffling gait, and skunk prints rarely overlap.

Eastern Spotted Skunk

fore

hind

Fore Print
Length: 1–1.3 in (2.5–3.3 cm)
Width: 0.9–1.1 in (2.3–2.8 cm)
Hind Print
Length: 1.2–1.5 in (3–3.8 cm)
Width: 0.9–1.1 in (2.3–2.8 cm)
Straddle
2–3 in (5–7.5 cm)
Stride
Walking: 1.5–3 in (3.8–7.5 cm)
Bounding: 6–12 in (15–30 cm)
Size
Length: 13–25 in (33–65 cm)
Weight
0.6–2.2 lb (0.3–1 kg)

walking   *bounding*

EASTERN SPOTTED SKUNK

Spilogale putorius

This beautifully marked skunk, smaller than its striped cousin, is found only in southern Pennsylvania. It enjoys diverse habitats—such as scrubland, forests and farmland—but it is a rare sight, because of its nocturnal habits and because it dens up in winter, coming out only on warmer nights.

This skunk leaves a very haphazard trail as it forages for food on the ground, and it occasionally climbs trees, which it does with ease. Long claws on the forefeet often register, and the palm and heel may leave defined pad marks. Although this skunk rarely runs, when it does so it may bound along, leaving groups of four prints, hind in front of fore. It sprays only when truly provoked, so its powerful odor is less frequently detected than that of the Striped Skunk (p. 62).

Similar Species: The Striped Skunk has larger prints and less scattered tracks with a shorter running stride (or it jumps); it does not climb trees.

Snowshoe Hare

fore

hind

Fore Print
Length: 2–3 in (5–7.5 cm)
Width: 1.5–2 in (3.8–5 cm)
Hind Print
Length: 4–6 in (10–15 cm)
Width: 2–3.5 in (5–9 cm)
Straddle
6–8 in (15–20 cm)
Stride
Hopping: 0.8–4.3 ft (25–130 cm)
Size
Length: 12–21 in (30–53 cm)
Weight
2–4 lb (0.9–1.8 kg)

hopping

SNOWSHOE HARE
(Varying Hare)
Lepus americanus

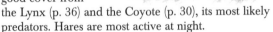

This hare is well known for its color change from summer brown to winter white and for its huge hind feet, which enable it to 'float' on top of snow. Found throughout this region, it frequents brushy areas in forests, which provide good cover from the Lynx (p. 36) and the Coyote (p. 30), its most likely predators. Hares are most active at night.

The Snowshoe Hare's most common track pattern is a hopping one, with triangular four-print groups; they can be quite long if the hare moves quickly. In winter, heavy fur on the hind feet (much larger than the forefeet) thickens the toes, which can splay out to further distribute the hare's weight on snow. Hares make well-worn runways that are often used as escape runs. You may encounter a resting hare, because hares do not live in burrows. Twigs and stems neatly cut at a 45° angle also indicate this hare's presence.

Similar Species: The Eastern Cottontail (p. 70) has much smaller prints. The European Hare (p. 68), which lives in New York, splays its hind toes less.

European Hare

fore

hind

Fore Print
Length: 2–4 in (5–10 cm)
Width: 2 in (5 cm)

Hind Print
Length: 5 in (13 cm)
Length with heel: 9 in (23 cm)
Width: 2.5 in (6.5 cm)

Straddle
8 in (20 cm)

Stride
Hopping: 1.5–3 ft (45–90 cm)
Bounding: to 12 ft (3.7 m)

Size
Length: 25–30 in (65–75 cm)

Weight
7–12 lb (3.2–5.5 kg)

hopping

EUROPEAN HARE
Lepus europaeus

This large, sturdy colonizer of open fields was introduced to North America over 100 years ago. It can now be found in much of New York. Unlike the much smaller Snowshoe Hare (p. 66), which turns white in winter, the European Hare remains brown, and it has a black spot on its tail.

The usual track pattern of this hare, typical for rabbits and hares, consists of elongated triangular clusters of prints. The forefeet register first (usually in a slightly diagonal line), leaving small, roundish prints that show four toes. The two much larger hind prints then fall (usually side by side) ahead of the fore prints. The hind prints are much longer when the whole heel registers, such as when the hare stops for a moment.

Similar Species: The smaller Eastern Cottontail (p. 70) has much smaller prints and a narrower straddle and stride. The hind toes of the Snowshoe Hare splay out more in snow.

Eastern Cottontail

fore

hind

Fore Print
Length: 1–1.5 in (2.5–3.8 cm)
Width: 0.8–1.3 in (2–3.3 cm)

Hind Print
Length: 3–3.5 in (7.5–9 cm)
Width: 1–1.5 in (2.5–3.8 cm)

Straddle
4–5 in (10–13 cm)

Stride
Hopping: 0.6–3 ft (18–90 cm)

Size
Length: 12–17 in (30–43 cm)

Weight
1.3–3 lb (0.6–1.4 kg)

hopping

EASTERN COTTONTAIL

Sylvilagus floridanus

This abundant rabbit is found throughout New York and Pennsylvania. Preferring brushy areas in grasslands and cultivated areas, it might be found in dense vegetation, where it hides from predators such as the Bobcat (p. 38) and the Coyote (p. 30). Largely nocturnal, the Eastern Cottontail might be seen at dawn or dusk and on darker days.

As with other rabbits and hares, this rabbit's most common track pattern is a triangular grouping of four prints, with the larger hind prints (which can appear pointed) falling in front of the fore prints (which may overlap). The hairiness of the toes will hide any pad detail. If you follow this rabbit's trail, you could be startled if it flies out from its 'form,' a depression in the ground in which it rests.

Similar Species: The Snowshoe Hare (p. 66) makes much larger prints, as does the European Hare (p. 68). The New England Cottontail (*S. transitionalis*) makes identical prints.

Porcupine

fore

hind

Fore Print
Length: 2.3–3.3 in (5.8–8.5 cm)
Width: 1.3–1.9 in (3.3–4.8 cm)
Hind Print
Length: 2.8–4 in (7–10 cm)
Width: 1.5–2 in (3.8–5 cm)
Straddle
5.5–9 in (14–23 cm)
Stride
Walking: 5–10 in (13–25 cm)
Size
Length with tail: 25–40 in (65–100 cm)
Weight
10–28 lb (4.5–13 kg)

walking

PORCUPINE
Erethizon dorsatum

This notorious rodent rarely runs—its many long quills are a formidable defense. Found throughout most of New York and Pennsylvania, the Porcupine prefers forests, but it can also be seen in more open areas.

The Porcupine's preferred pigeon-toed, waddling gait leaves an alternating track pattern, with the hind print registered on or slightly in front of the shorter fore print. Look for long claw marks on all prints. The fore print shows four toes, and the hind print shows five. Clear prints may show the unusual pebbly surface of the solid heel pads, but a Porcupine's tracks are often scratch-marked by its heavy, spiny tail. In deeper snow this squat animal drags its feet, and it may leave a trough with its body. A Porcupine's trail might lead you to a tree, where this animal spends much of its time feeding; if so, look for chewed bark or nipped twigs on the ground.

Similar Species: The Badger (p. 60) makes pigeon-toed prints, but its tracks don't show tail drag, and it doesn't climb trees.

Beaver

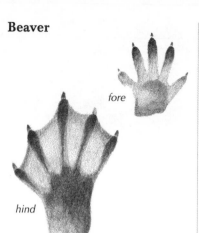

fore

hind

Fore Print
Length: 2.5–4 in (6.5–10 cm)
Width: 2–3.5 in (5–9 cm)

Hind Print
Length: 5–7 in (13–18 cm)
Width: 3.3–5.3 in (8.5–13 cm)

Straddle
6–11 in (15–28 cm)

Stride
Walking: 3–6.5 in (7.5–17 cm)

Size
Length with tail: 3–4 ft (90–120 cm)

Weight
28–75 lb (13–34 kg)

walking

BEAVER
Castor canadensis

Few animals leave as many signs of their presence as the Beaver, North America's largest rodent and a common sight around water. Look for the conspicuous dams and lodges—capable of changing the local landscape—and the stumps of felled trees. Inspect trunks gnawed clean of bark for marks of the Beaver's huge incisors. Scent mounds marked with castoreum, a strong-smelling yellowish fluid that Beavers produce, also indicate recent activity.

Check the large hind prints for signs of webbing and broad toenails. The nail of the fourth toe usually does not register, and it is rare for all five toes on each foot to do so. Irregular foot placement in the alternating walking gait may produce a direct register or a double register. The Beaver's thick, scaly tail may mar its tracks, as can the branches that it drags about for construction and food. Repeated path use results in well-worn trails.

Similar Species: The Beaver's many signs, including its large hind prints, minimize confusion. Muskrat (p. 76) prints are smaller.

Muskrat

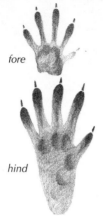

fore

hind

Fore Print
Length: 1.1–1.5 in (2.8–3.8 cm)
Width: 1.1–1.5 in (2.8–3.8 cm)

Hind Print
Length: 1.6–3.2 in (4–8 cm)
Width: 1.5–2.1 in (3.8–5.3 cm)

Straddle
3–5 in (7.5–13 cm)

Stride
Walking: 3–5 in (7.5–13 cm)
Running: to 1 ft (30 cm)

Size
Length with tail: 16–25 in (40–65 cm)

Weight
2–4 lb (0.9–1.8 kg)

walking

MUSKRAT
Ondatra zibethicus

Like the Beaver (p. 74), this rodent can be found throughout New York and Pennsylvania, wherever there is water. Beavers are very tolerant of Muskrats and even allow them to live in parts of their lodges. Active all year, the Muskrat leaves plenty of signs. It digs extensive networks of burrows, often undermining riverbanks, so do not be surprised if you suddenly fall into a hidden hole! Also look for small lodges in the water and beds of vegetation on which the Muskrat rests, suns and feeds in summer.

The small fifth (inner) toe of the forefoot rarely registers. Stiff hairs that aid in swimming may create a 'shelf' around the five well-formed toes of the hind print. The common alternating walking pattern shows print pairs that alternate from side to side; the hind print is just behind the fore print or slightly overlaps it. In snow, a Muskrat's feet drag, and its tail leaves a sweeping dragline.

Similar Species: Few animals share this water-loving rodent's habits. The Beaver makes larger tracks and leaves many other signs.

Woodchuck

fore

hind

Fore and Hind Prints
Length: 1.8–2.8 in (4.5–7 cm)
Width: 1–2 in (2.5–5 cm)

Straddle
3.3–6 in (8.5–15 cm)

Stride
Walking: 2–6 in (5–15 cm)
Bounding: 6–14 in (15–35 cm)

Size (male>female)
Length with tail:
20–25 in (50–65 cm)

Weight
5.5–12 lb (2.5–5.5 kg)

walking   *bounding*

WOODCHUCK
(Whistle Pig, Groundhog, Marmot)

Marmota monax

This robust member of the squirrel family is a common sight in open woodlands and adjacent open areas throughout New York and Pennsylvania. Always on the watch for predators, but not too troubled by humans, the Woodchuck never wanders far from its burrow. This marmot hibernates during winter but emerges in early spring; look for tracks in late spring snowfalls and in mud around the burrow entrances.

A Woodchuck's fore print shows four toes, three palm pads and two heel pads (not always evident). The hind print shows five toes, four palm pads and two poorly registering heel pads. The Woodchuck usually leaves an alternating walking pattern, with the hind print registered on the fore print. When a Woodchuck runs from danger, it makes groups of four prints, hind ahead of fore.

Similar Species: Squirrel (pp. 82–89) prints are very similar but smaller. A small Raccoon's (p. 42) bounding track pattern will be similar, but it will show five-toed fore prints.

Eastern Chipmunk

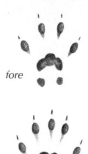

fore

hind

Fore Print
Length: 0.8–1 in (2–2.5 cm)
Width: 0.4–0.8 in (1–2 cm)

Hind Print
Length: 0.7–1.3 in (1.8–3.3 cm)
Width: 0.5–0.9 in (1.3–2.3 cm)

Straddle
2–3.2 in (5–8 cm)

Stride
Bounding: 7–15 in (18–38 cm)

Size
Length with tail: 7–10 in (18–25 cm)

Weight
2.5–5 oz (70–140 g)

bounding

EASTERN CHIPMUNK
Tamias striatus

Look for this delightful character throughout New York and Pennsylvania. The Eastern Chipmunk is found in a variety of habitats, from the dense forest floor to open areas near buildings. You are more likely to see or hear this rodent, which is highly active during summer, than to notice its tracks. This large chipmunk is happiest on the ground, but it will gladly climb sturdy oak trees to harvest juicy, ripe acorns. It enters a deep sleep in winter, waking up from time to time to eat.

Chipmunks are so light that their tracks rarely show fine details. The forefeet each have four toes and the hind feet have five. Chipmunks run on their toes, so the two heel pads of the forefeet seldom register; the hind feet have no heel pads. Their erratic track patterns, like those of many of their cousins, show the hind feet registered in front of the forefeet. A chipmunk trail often leads to extensive burrows.

Similar Species: No other chipmunks inhabit this region. Squirrels (pp. 82–89) usually have larger prints and a wider straddle, and they are more likely to make midwinter tracks. Mouse (pp. 96–99) tracks are smaller.

Eastern Gray Squirrel

fore

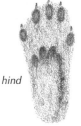

hind

bounding

Fore Print
Length: 1–1.8 in (2.5–4.5 cm)
Width: 1 in (2.5 cm)
Hind Print
Length: 2.3–3 in (5.8–7.5 cm)
Width: 1.1–1.5 in (2.8–3.8 cm)
Straddle
3.8–6 in (9.5–15 cm)
Stride
Bounding: 0.7–3 ft (22–90 cm)
Size
Length with tail: 17–20 in (43–50 cm)
Weight
14–25 oz (400–710 g)

EASTERN GRAY SQUIRREL

Sciurus carolinensis

This large and familiar squirrel can be a common sight in deciduous and mixed forests throughout these two states, even in urban areas. Active all year, the Eastern Gray Squirrel can leave a wealth of evidence, especially in winter as it scurries about digging up nuts that it buried during the previous fall.

The Eastern Gray Squirrel leaves a typical squirrel track pattern when it runs or bounds. The hind prints fall slightly in front of the fore prints. A clear fore print shows four toes with sharp claws, four fused palm pads and two heel pads. The hind print shows five toes and four palm pads; if the full heel-length registers, it also shows two small heel pads.

Similar Species: Fox Squirrel (p. 84) prints are as big or larger. Red Squirrel (p. 86) prints are smaller. Chipmunks (p. 80) and flying squirrels (p. 88) make smaller tracks in a similar pattern but with narrower straddles. Rabbits and hares (pp. 66–71) make longer track patterns, and their forefeet rarely register side by side when they run.

Fox Squirrel

fore

hind

Fore Print
Length: 1–1.9 in (2.5–4.5 cm)
Width: 1–1.7 in (2.5–4.3 cm)
Hind Print
Length: 2–3.3 in (5–7.5 cm)
Width: 1.5–1.9 in (3.8–4.5 cm)
Straddle
4–6 in (10–15 cm)
Stride
Bounding: 0.7–3 ft (22–90 cm)
Size
Length with tail: 18–28 in (45–70 cm)
Weight
1–2.4 lb (0.5–1.1 kg)

bounding

FOX SQUIRREL
Sciurus niger

This squirrel is much like the Eastern Gray Squirrel (p. 82), but larger and with a yellowish underside. It can be a common sight in deciduous forests with plenty of nut trees and in open areas or woodlands of southern New York and Pennsylvania. Piles of nutshells at tree bases indicate its favorite feeding sites. Active all year, the Fox Squirrel spends a lot of time foraging on the ground, often collecting nuts that it buried singly during the previous fall.

A clear fore print shows four toes with claws evident, four fused palm pads and two heel pads. The hind print shows five toes, four palm pads and sometimes a heel. When it runs or bounds, the Fox Squirrel makes a typical squirrel track, the hind prints slightly in front of the fore prints, and the prints in each pair roughly side by side.

Similar Species: Eastern Gray Squirrel prints are generally smaller. Red Squirrel (p. 86) prints are much smaller. Chipmunk (p. 80) and flying squirrel (p. 88) tracks, which are in a similar pattern, are smaller and have narrower straddles. Rabbits and hares (pp. 66–71) make longer track patterns, and their forefeet rarely register side by side when they run.

Red Squirrel

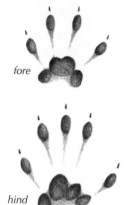

fore

hind

Fore Print
Length: 0.8–1.5 in (2–3.8 cm)
Width: 0.5–1 in (1.3–2.5 cm)

Hind Print
Length: 1.5–2.3 in (3.8–5.8 cm)
Width: 0.8–1.3 in (2–3.3 cm)

Straddle
3–4.5 in (7.5–11 cm)

Stride
Bounding: 8–30 in (20–75 cm)

Size
Length with tail:
    9–15 in (23–38 cm)

Weight
2–9 oz (55–260 g)

bounding

bounding
(deep snow)

RED SQUIRREL
(Pine Squirrel, Chickaree)
Tamiasciurus hudsonicus

When you
enter a Red
Squirrel's terri-
tory, the inhabi-
tant greets you with
a loud, chattering call.
Another obvious sign of this
forest dweller, which is found
in forested areas of the two states,
is its large middens—piles of cone scales and cores
left beneath trees—that indicate favorite feeding sites.

Active all year in its small territory, a Red Squirrel
will leave an abundance of trails that lead from tree to
tree or down a burrow. This energetic animal mostly
bounds, leaving four-print groups, the hind prints falling
in front of the fore prints (which tend to be side by side,
but not always). Four toes show on each fore print, and
five on each hind print. The heels often do not register
when a squirrel moves quickly. In deeper snow the prints
merge to form pairs of diamond-shaped tracks.

Similar Species: Fox Squirrel (p. 84) and Eastern Gray
Squirrel (p. 82) prints are larger. Chipmunk (p. 80) and
flying squirrel (p. 88) tracks are in a similar pattern, but
they are smaller and with a narrower straddle.

Southern Flying Squirrel

fore

hind

Fore Print
Length: 0.3–0.5 in (0.8–1.3 cm)
Width: 0.4 in (1 cm)

Hind Print
Length: 0.9–1.3 in (2.3–3.3 cm)
Width: 0.5 in (1.4 cm)

Straddle
2–2.5 in (5–6.5 cm)

Stride
Bounding: 7–22 in (18–55 cm)

Size
Length with tail: 8–10 in (20–25 cm)

Weight
1.5–3.2 oz (43–90 g)

bounding (in snow)

SOUTHERN FLYING SQUIRREL
Glaucomys volans

This soft-furred brown acrobat is capable of long-distance gliding using the glide membranes that connect its forelegs and hind legs. Primarily nocturnal, the Southern Flying Squirrel is found in coniferous and mixed forests throughout the region. In winter, up to 50 Southern Flying Squirrels can be found huddled together in a nest for warmth. Like other tree squirrels, this one does not truly hibernate.

Flying squirrels spend so much of their time in trees and gliding in the air that they make few tracks. If you do encounter tracks, they will start suddenly with a sitz-mark and continue immediately to the nearest tree. New information from Mark Elbroch, an East Coast tracking expert, shows that the characteristic pattern for this squirrel is a bound where the front tracks register in front of the rear tracks, and the straddle is narrower than that of other squirrels.

Similar Species: Chipmunk (p. 80) tracks are very similar. Other squirrels (pp. 82–87) usually make larger prints and rarely leave sitzmarks, but with tracks in deep snow it can be impossible to identify the squirrel species.

Appalachian Woodrat

fore

hind

Fore Print
Length: 0.6–0.8 in (1.5–2 cm)
Width: 0.4–0.5 in (1–1.3 cm)

Hind Print
Length: 1–1.5 in (2.5–3.8 cm)
Width: 0.6–0.8 in (1.5–2 cm)

Straddle
2.3–2.8 in (5.8–7 cm)

Stride
Walking: 1.8–3 in (4.5–7.5 cm)
Bounding: 5–8 in (13–20 cm)

Size
Length with tail:
    14–17 in (35–43 cm)

Weight
13–16 oz (370–450 g)

walking   *bounding*

APPALACHIAN WOODRAT
(Allegheny Woodrat)
Neotoma magister

This nocturnal woodrat lives in rocky areas of much of Pennsylvania. The trail of this woodrat might lead you to its distinctive mass of a nest, most often under a rock or in a crevice. Although it favors rocky areas, it feeds on foliage, seeds, ferns and fungi and therefore requires nearby vegetation.

Four toes show on the fore print and five on the hind. The short claws rarely register. A woodrat often walks in an alternating fashion, with the hind print direct registering on the fore print. This woodrat frequently bounds as well, leaving a pattern of four prints, with the larger hind print in front of the diagonally placed fore prints. The stride tends to be short relative to the size of the prints.

Similar Species: Indistinct Red Squirrel (p. 86) prints are similar but usually larger. The Norway Rat (p. 92) has similar prints, but it is usually found close to human activity. Woodchuck (p. 78) prints are similar but much larger.

Norway Rat

fore

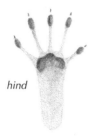

hind

Fore Print
Length: 0.7–0.8 in (1.8–2 cm)
Width: 0.5–0.7 in (1.3–1.8 cm)

Hind Print
Length: 1–1.3 in (2.5–3.3 cm)
Width: 0.8–1 in (2–2.5 cm)

Straddle
2–3 in (5–7.5 cm)

Stride
Walking: 1.5–3.5 in (3.8–9 cm)
Bounding: 9–20 in (23–50 cm)

Size
Length with tail: 13–19 in (33–48 cm)

Weight
7–18 oz (200–510 g)

walking

NORWAY RAT
(Brown Rat)
Rattus norvegicus

Active both day
and night, this despised rat
is widespread almost anywhere
that humans have decided to build
their homes. Not entirely dependent
on people, it may live in the wild as well.

The fore print shows four toes, and the hind print
shows five. When it bounds, this colonial rat leaves four-
print groups, with the hind prints in front of the diagonally placed fore prints. Sometimes one of the hind feet
direct registers on a fore print, creating a three-print
group. This rat more commonly leaves an alternating
walking pattern with the larger hind prints close to or
overlapping the fore prints; the hind heel does not show.
The tail often leaves a dragline in snow. Rats live in
groups, so you may find many trails together, often
leading to their 5-inch (2-cm) wide burrows.

Similar Species: Mouse (pp. 96–99) prints are much
smaller, with different track patterns. Chipmunk (p. 80)
tracks are similar but smaller, and the track patterns differ. Red Squirrel (p. 86) tracks show distinctive squirrel
traits.

93

Woodland Vole

fore

hind

Fore Print
Length: 0.5 in (1.3 cm)
Width: 0.5 in (1.3 cm)

Hind Print
Length: 0.6 in (1.5 cm)
Width: 0.5–0.8 in (1.3–2 cm)

Straddle
1.3–2 in (3.3–5 cm)

Stride
Walking/Trotting: 0.8 in (2 cm)
Bounding: 2–6 in (5–15 cm)

Size
Length with tail:
    4–5.5 in (10–14 cm)

Weight
0.8–1.3 oz (23–37 g)

walking

bounding (in snow)

WOODLAND VOLE
(Pine Vole)

Microtus pinetorum

Though a
positive identifi-
cation of a vole track
can be challenging, one
of the most likely candidates
in New York and Pennsylvania
is the Woodland Vole. It lives in pine
forests and just about every other habitat as well.

When clear (which is seldom), vole fore prints show
four toes, and hind prints show five. A vole's walk and
trot both leave a paired alternating track pattern with a
hind print occasionally direct registered on a fore print.
Voles usually opt for a faster bounding; the resulting
print pairs show the hind prints registered on the fore
prints. These voles lope quickly across open areas, creating a three-print pattern. In winter, voles stay under the
snow; when it melts, look for distinctive piles of cut grass
from their ground nests. The bark at the bases of shrubs
may show tiny teeth marks left by gnawing. In summer,
well-used vole paths appear as little runways in the grass.

Similar Species: Other common voles in the region
include the Southern Red-backed Vole (*Cletherionomys
gapperi*) and the Meadow Vole (*M. pennsylvanicus*). Mouse
(pp. 96–99) bounding tracks show four-print groups.

Deer Mouse

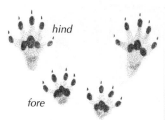

hind

fore

bounding group

Fore Print
Length: 0.3–0.4 in (0.8–1 cm)
Width: 0.3–0.4 in (0.8–1 cm)
Hind Print
Length: 0.3–0.5 in (0.8–1.3 cm)
Width: 0.3–0.4 in (0.8–1 cm)
Straddle
1.4–1.8 in (3.5–4.5 cm)
Stride
Bounding: 5–12 in (13–30 cm)
Size
Length with tail:
    6–12 in (15–30 cm)
Weight
0.5–1.3oz (14–35 g)

bounding

bounding
(in snow)

DEER MOUSE
Peromyscus maniculatus

The highly adaptable Deer Mouse—one of the region's most abundant mammals—lives anywhere from arid valleys all the way up to alpine meadows. It is seldom seen, because it is nocturnal. The Deer Mouse may enter buildings in winter, where it will stay active.

In perfect, soft mud, the fore prints each show four toes, three palm pads and two heel pads, and the hind prints show five toes and three palm pads; the heel pads rarely register. Bounding tracks, most noticeable in snow, show the hind prints falling in front of the fore prints. In soft snow the prints may merge to look like larger pairs of prints; tail drag will be evident. A mouse trail may lead up a tree or down into a burrow.

Similar Species: Many less common species of mice have near-identical tracks. The House Mouse (*Mus musculus*), with very similar tracks, associates more with humans. Jumping mouse (p. 98) prints show long, thin toes. Voles (p. 94) tend to trot and have a much shorter bounding track pattern. Chipmunks (p. 80) have a wider straddle. Shrews (p. 100) have a narrower straddle.

Meadow Jumping Mouse

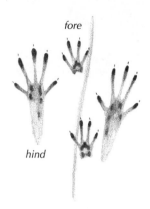

fore

hind

Fore Print
Length: 0.3–0.5 in (0.8–1.3 cm)
Width: 0.3–0.5 in (0.8–1.3 cm)
Hind Print
Length: 0.5–1.3 in (1.3–3.3 cm)
Width: 0.5–0.7 in (1.3–1.8 cm)
Straddle
1.8–1.9 in (4.5–4.8 cm)
Stride
Bounding: 7–18 in (18–45 cm)
In alarm: 3–6 ft (90–180 cm)
Size
Length with tail: 7–9 in (18–23 cm)
Weight
0.6–1.3 oz (17–35 g)

bounding

MEADOW JUMPING MOUSE
Zapus hudsonius

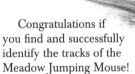

Congratulations if
you find and successfully
identify the tracks of the
Meadow Jumping Mouse!
Though it is abundant throughout the region, its preference for grassy meadows and dense undergrowth—and
its long, deep winter hibernation (about six months!)—
makes locating tracks very difficult.

Jumping mouse tracks are distinctive if you do find
them. The two smaller forefeet register between the long
hind prints; the long heels do not always register, and
some prints show just the three long middle toes. The
toes on the forefeet may splay so much that the side toes
point backward. When they bound, jumping mice make
short leaps. The tail may leave a dragline in soft mud
or unseasonable snow. Clusters of cut grass stems about
5 inches (13 cm) long and lying in meadows are a more
abundant sign of this rodent.

Similar Species: Deer Mouse (p. 96) tracks may have
the same straddle. Heel-less hind prints may be mistaken
for a vole's prints (p. 94), or a small bird's (pp. 124–127)
or even an amphibian's (pp. 128–133).

Masked Shrew

hind

fore

bounding group

Fore Print
Length: 0.2 in (0.5 cm)
Width: 0.2 in (0.5 cm)
Hind Print
Length: 0.6 in (1.5 cm)
Width: 0.3 in (0.8 cm)
Straddle
0.8–1.3 in (2–3.3 cm)
Stride
Bounding: 1.2–3 in (3–7.5 cm)
Size
Length with tail: 2.3–4.5 in (7–11 cm)
Weight
0.1–0.3 oz (3–9 g)

bounding

MASKED SHREW
Sorex cinereus

Though several species of tiny, frenetic shrews are found in this region, the widespread and adaptable Masked Shrew is a likely candidate if you find tracks. This small shrew prefers moist fields, marshes, bogs or woodlands, but it can also be found in higher and drier grasslands.

In its energetic and unending quest for food, a shrew usually leaves a four-print bounding pattern, but it may slow to an alternating walking gait. The individual prints in a group are often indistinct, but in mud or shallow, wet snow you can even count the five toes on each print. In deeper snow a shrew's tail often leaves a dragline. If a shrew tunnels under the snow, it may leave a snow ridge on the surface. A shrew's trail may disappear down a burrow.

Similar Species: The Long-tailed (Rock) Shrew (*S. dispar*) and the Smoky Shrew (*S. fumeus*) both make indistinguishable prints. The Pygmy Shrew (*S. hoyi*), also widespread, has slightly smaller prints. A mouse's (pp. 96–99) fore prints will show four toes.

Star-nosed Mole

a molehill of the Star-nosed Mole

Size
Length with tail: 6–8.5 in (15–22 cm)
Weight
1–2.6 oz (28–75 g)

STAR-NOSED MOLE
Condylura cristata

This peculiar character
may be found throughout the
two states. The Star-nosed Mole is
easily identified by the strange tentacle-like protrusions
on its nose, thought to help it find food in its dark and
often subterranean world. It spends more time out of
its burrows than other moles do, usually at night.

Typical signs of this mole include the big piles of soil
pushed out of its burrows. Because of its preference for
swimming and wet areas, look for this mole's hills along
the banks of streams and rivers and in raised areas
around marshes and wet fields. Clear prints from its
long-clawed feet are rarely found. The Star-nosed Mole
remains under the snow during winter and swims under
the ice, so even winter tracks are seldom seen.

Similar Species: The Hairy-tailed Mole (*Parascalops
breweri*) spends more time in its burrow, but it may leave
similar signs, usually not as close to water.

104

BIRDS, AMPHIBIANS & REPTILES

A guide to the animal tracks found in New York and Pennsylvania is not complete without some consideration of the birds, amphibians and reptiles found in the region.

Several bird species have been chosen to represent the main types common to these two states, but remember that individual bird species are not easily identified by track alone. Bird tracks are often abundant in snow and are clearest in shallow, wet snow. The shores of lakes and streams are very reliable places to find bird tracks—the mud there can hold a clear print for a long time. The sheer number of tracks made by shorebirds and waterfowl can be astonishing. Though some bird species prefer to perch in trees or soar across the sky, it can be entertaining to follow the tracks of birds that spend a lot of time on the ground. They can spin around in circles and lead you in all directions. The trail may suddenly end as the bird takes flight, or it might terminate in a pile of feathers, the bird having fallen victim to a predator.

Many amphibians and turtles depend on moist environments, so look in the soft mud along the shores of lakes and ponds for their distinctive tracks. You may be able to distinguish frog tracks from toad tracks, because these amphibians generally move differently, but it can be very difficult to identify the species. Reptiles thrive and outnumber the amphibians in drier environments, but they seldom leave good tracks, except in occasional mud or perhaps in sand. Snakes leave distinctive body prints.

Mallard

Print
Length: 2–2.5 in (5–6.5 cm)
Straddle
4 in (10 cm)
Stride
to 4 in (10 cm)
Size
23 in (58 cm)

MALLARD
Anas platyrhynchos

male

female

This dabbling duck—the male a familiar sight with its striking green head—is common in open areas near lakes and ponds. Its webbed feet leave prints that can often be seen in abundance along the muddy shores of just about any waterbody, including those in urban parks.

The webbed foot of the Mallard has three long toes that all point forward. Though the toes register well, the webbing between the toes does not always show in the print. The Mallard's inward-pointing feet give it a pigeon-toed appearance and perhaps account for its waddling gait, a characteristic for which ducks are known.

Similar Species: Many waterfowl, such as other ducks, as well as the Herring Gull (p. 108), leave similar prints. Exceptionally large prints were likely made by geese (various species) or by a swan (*Cygnus* spp.).

Herring Gull

Print
Length: 3.5 in (9 cm)
Straddle
4–6 in (10–15 cm)
Stride
4.5 in (11 cm)
Size
Length: 23–25 in (58–65 cm)

HERRING GULL
Larus argentatus

The Herring Gull, with its long wings and webbed toes, is a strong long-distance flier as well as an excellent swimmer. This bird is increasingly common in New York and Pennsylvania. It is concentrated in great numbers near waterbodies and garbage dumps.

Gulls leave slightly asymmetrical tracks that show three toes. They have claws that register outside the webbing, and the claw marks are usually attached to the footprint. Most gulls have quite a swagger to their gait, and they leave a trail with the tracks turned strongly inward.

Similar Species: Gull species cannot be reliably identified by track alone, but smaller species have conspicuously smaller tracks. Mallard (p. 106) and other duck tracks are often difficult to distinguish from gull tracks. Geese (various species) make larger prints.

Great Blue Heron

Print
Length: to 6.5 in (17 cm)
Straddle
8 in (20 cm)
Stride
9 in (23 cm)
Size
4.2–4.5 ft (1.3–1.4 m)

GREAT BLUE HERON
Ardea herodias

The regal and graceful image of this large heron symbolizes the precious wetlands in which it patiently hunts for food. Usually still and statuesque as it waits for a meal to swim by, this heron will have cause to walk from time to time, perhaps to find a better hunting location. Look for its large, slender tracks along the banks or mudflats of waterbodies.

Not surprisingly, a bird that lives and hunts with such precision walks in a similar fashion, leaving straight tracks that fall in a nearly straight line. Look for the slender rear toe in the print.

Similar Species: The American Bittern (*Botaurus lentiginosus*) and the Green Heron (*Butorides virescens*) make similar but smaller tracks. Cranes (*Grus* spp.), which occupy similar habitats but are rarely seen in the region, leave similarly sized prints, but a crane's rear toes are smaller and do not register.

Common Snipe

Print
Length: 1.5 in (3.8 cm)
Straddle
to 1.8 in (4.5 cm)
Stride
to 1.3 in (3.3 cm)
Size
11–12 in (28–30 cm)

COMMON SNIPE
Gallinago gallinago

This short-legged character is a resident of marshes and bogs, where its neat prints can often be seen in mud. Snipes are quite secretive when on the ground, and so you may be surprised if one suddenly flushes out from beneath your feet. If there is a Common Snipe in the air, you may hear an eerie whistle if it dives from the sky.

The Common Snipe's neat prints show four toes, including a small rear toe that points inward. The bird's short legs and stocky body give it a very short stride.

Similar Species: Many shorebirds, including the Spotted Sandpiper (p. 114), leave similar tracks.

Spotted Sandpiper

Print
Length: 0.8–1.3 in (2–3.3 cm)
Straddle
to 1.5 in (3.8 cm)
Stride
Erratic
Size
7–8 in (18–20 cm)

SPOTTED SANDPIPER
Actitis macularia

The bobbing tail of the Spotted Sandpiper is a common sight on the shores of lakes, rivers and streams, but you will usually find just one of these territorial birds in any given location. Because of its excellent camouflage, likely the first that you will see of this bird will be when it flies away, its fluttering wings close to the surface of the water.

As it teeters up and down on the shore, a sandpiper leaves trails of three-toed prints. Its fourth toe is very small and faces off to one side at an angle. Sandpiper tracks often have an erratic stride.

Similar Species: All sandpipers and plovers, including the common Killdeer (*Charadrius vociferus*), leave similar tracks, although there is much diversity in size. The Common Snipe (p. 112) makes similar but larger tracks.

Ruffed Grouse

Print
Length: 2–3 in (5–7.5 cm)
Straddle
0.5–2 in (1.3–5 cm)
Stride
Walking: 3–6 in (7.5–15 cm)
Size
15–19 in (38–48 cm)

RUFFED GROUSE

Bonasa umbellus

This ground-dweller prefers the quiet seclusion of coniferous forests, where its excellent camouflage usually affords it good protection. Although you might find its tracks in mud, snow much improves your chances of finding them. If you follow a Ruffed Grouse trail quietly, you may be startled when the bird bursts from cover almost beneath your feet.

The three thick front toes leave very clear impressions, but the short rear toe, which is angled off to one side, does not always show up so well. This bird's neat, straight trail appears to reflect its cautious approach to life on the forest floor.

Similar Species: The Spruce Grouse (*Dendragapus canadensis*) in northern New York leaves similar tracks. The Wild Turkey (*Meleagris gallopavo*) leaves similar but much larger tracks.

Great Horned Owl

Strike
Width: to 3 ft (90 cm)
Size
22 in (55 cm)

GREAT HORNED OWL

Bubo virginianus

Often seen resting quietly in trees by day, this wide-ranging owl prefers to hunt at night. You might find an untidy hole in the snow, possibly surrounded by wing and tail-feather imprints—a well-registered 'strike' can be quite a sight. The Great Horned Owl strikes through the snow with its talons, and the feather imprints are made as the owl struggles to take off with possibly heavy prey. An ungraceful walker, it prefers to fly away from the scene, though its tracks may be evident near roadkill.

You might stumble across a strike and guess that the owl's target could have been a vole (p. 94) scurrying around underneath the snow. Or you may be following the surface trail of an animal to find that it abruptly ends with this strike mark, where the animal has been seized.

Similar Species: If the prey left no approaching trail on the snow, the strike is likely an owl's, because owls hunt by sound. Otherwise, the strike, usually with less rounded, more distinct feather imprints, could be by a hawk or a Common Raven (*Corvus corax*), both of which hunt by sight.

American Crow

Print
Length: 2.5–3 in (6.5–7.5 cm)
Straddle
1.5–3 in (3.8–7.5 cm)
Stride
Walking: 4 in (10 cm)
Size
16 in (40 cm)

AMERICAN CROW
Corvus brachyrhyncos

The black silhouette of the American Crow can be a common sight in a variety of habitats. A crow will frequently come down to the ground and contentedly strut around. Its loud *caw* can be heard from quite a distance. Crows can be especially noisy when they are mobbing an owl or a hawk.

The American Crow typically leaves an alternating walking track pattern. Its prints show three sturdy toes pointing forward and one toe pointing backward. When a crow is in need of greater speed, perhaps for take-off, it bounds along leaving irregular pairs of diagonally placed prints with a longer stride between each pair.

Similar Species: Other corvids, such as jays (various species), also spend a lot of time on the ground and make similar tracks that vary in size according to the size of the bird. The much larger Common Raven (*C. corax*) leaves tracks to 4 inches (10 cm) long, with a stride of 6 inches (15 cm).

Northern Flicker

Print
Length: 1.8 in (4.5 cm)
Straddle
1–1.5 in (2.5–3.8 cm)
Stride
Hopping: 1.5–5 in (3.8–13 cm)
Size
5.5–6.5 in (14–17 cm)

NORTHERN FLICKER

Colaptes auratus

male

female

This attractive woodpecker, which can be seen throughout these two states, is common in open woodlands and right into suburban areas. Unlike most woodpeckers, it spends some of its time feeding on the ground.

A clear flicker track shows a distinctive arrangement of two strong toes pointing forward and two pointing to the rear, with the outer toes slightly longer than the inner ones. The flicker's toes—along with short, strong legs that give the bird a short stride—are well suited for grasping tree trunks and limbs as this agile bird works its way along in search of insects.

Similar Species: Most other birds have very different tracks. Other woodpeckers would leave similar tracks, but very few come down to the ground as much as the obliging Northern Flicker does.

Dark-eyed Junco

Print
Length: to 1.5 in (3.8 cm)
Straddle
1–1.5 in (2.5–3.8 cm)
Stride
Hopping: 1.5–5 in (3.8–13 cm)
Size
5.5–6.5 in (14–17 cm)

DARK-EYED JUNCO
Junco hyemalis

This common small bird typifies the many small hopping birds found in the region. Each foot has three forward-pointing toes and one longer toe at the rear. The best prints are left in snow, although in deep snow the toe detail is lost; the footprints may show some dragging between the hops.

A good place to study this type of prints is near a birdfeeder. Watch the birds scurry around as they pick up fallen seeds, then have a look at the prints left behind. For example, juncos are attracted to small seeds that chickadees (*Poecile* spp.) scatter as they forage for sunflower seeds in the birdfeeder. Also look for tracks under coniferous trees, where juncos feed on fallen seeds in winter.

Similar Species: The changing seasons influence bird diversity. Toe size may help with identification—larger birds make larger prints. The Northern Cardinal (p. 126) has very similar tracks. In powdery snow, junco tracks can resemble mouse (pp. 96–99) tracks, so follow the trail to see if it disappears down a hole or into thin air.

Northern Cardinal

Print
Length: to 1.5 in (3.8 cm)
Straddle
1–1.5 in (2.5–3.8 cm)
Stride
Hopping: 1.5–5 in (3.8–13 cm)
Size
9 in (23 cm)

NORTHERN CARDINAL
Cardinalis cardinalis

female

male

     The brilliant red plumage and small black mask and chin of the male are a joy to see on this year-round resident. Like the Dark-eyed Junco (p. 124), the Northern Cardinal is a small hopping bird.

Each foot has three forward-pointing toes and one longer toe at the rear. The best prints are left in snow, although in deep snow the toe detail is lost, and the feet may show some dragging between the hops.

A good place to study these types of prints is near a birdfeeder. Watch the birds scurry around as they pick up fallen seeds, then have a look at the prints that they have left behind.

Similar Species: Juncos, finches (various species) and sparrows (various species) make similar tracks. The size of the toes may indicate what kind of bird you are tracking—larger birds have larger footprints. Not all birds are present at all times of the year, so keep in mind the season when tracking.

Frogs

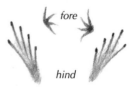

fore

hind

Straddle
to 3 in (7.5 cm)

hopping

FROGS

Wood
Frog

The smallest frogs include the treefrogs. The Spring Peeper (*Pseudacris crucifer*) is only 1.5 inches (3.8 cm) long and prefers thick undergrowth and shrubs near the water, so its tracks are a rare sight. The Gray Treefrog (*Hyla versicolor*) spends most of its time in trees, coming down to breed and sing at night. The Green Frog (*Rana clamitans*) and the Pickerel Frog (*R. palustris*), both widespread, favor slow-moving, shallow water and swampy areas. The beautiful, widespread Wood Frog (*R. sylvatica*), up to 3 inches (8 cm) long, inhabits moist woodland areas. Unusually large tracks are surely by the robust Bullfrog (*R. catesbeiana*). North America's largest frog, it can be 8 inches (20 cm) long.

The best place to look for frog tracks is along the muddy fringes of waterbodies. A frog's hopping action results in its two small forefeet registering in front of its long-toed hind prints. Frog tracks vary greatly in size, depending on species and age. Toads (p. 130) usually walk, but they may also hop.

Toads

hind   *fore*

Straddle
to 2.5 in (6.5 cm)

walking

TOADS

American Toad

There are fewer toad species than frog species in New York and Pennsylvania. The most widespread toad, and the one most likely to be encountered, is the American Toad (*Bufo americanus*), which lives in many different moist habitats. Fowler's Toad (*B. fowleri*) can be found in the south of New York State and in most of Pennsylvania in temporary pools and ditches. The Eastern Spadefoot (*Scaphiopus holbrookii*), found in eastern Pennsylvania and southern New York, is common in arid and semi-arid habitats. Toads in this region can grow up to 5 inches (12 cm) long.

Undoubtedly, the best place to look for toad tracks is, as with frog tracks, along the muddy fringes of water-bodies, but they can occasionally be found in drier areas—for example, as unclear trails in dusty patches of soil. In general, toads walk and frogs (p. 128) hop, but toads are pretty capable hoppers, too, especially when being hassled by overly enthusiastic naturalists. Toads leave rather abstract prints as they walk. The heels of the hind feet do not register. On less firm surfaces, the toes often leave draglines.

Salamanders

fore

hind

Straddle
to 3 in (7.5 cm)

walking

SALAMANDERS

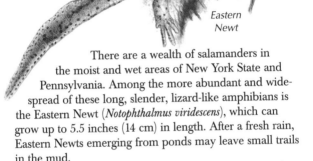

Eastern
Newt

There are a wealth of salamanders in
the moist and wet areas of New York State and
Pennsylvania. Among the more abundant and widespread of these long, slender, lizard-like amphibians is
the Eastern Newt (*Notophthalmus viridescens*), which can
grow up to 5.5 inches (14 cm) in length. After a fresh rain,
Eastern Newts emerging from ponds may leave small trails
in the mud.

The Spotted Salamander (*Ambystoma maculatum*),
which can grow to 10 inches (25 cm) long, lives throughout the region in mixed forests and some coniferous
forests. The smaller Jefferson Salamander (*Ambystoma
jeffersonianum*), with light blue spots on its undersides,
grows to 8 inches (21 cm) and is common under debris
near swamps and ponds in deciduous forests.

In general, a salamander's fore print shows four toes,
and the larger hind print shows five. However, print
detail is often blurred by the animal's dragging belly
or by the swinging of its thick tail across the tracks.

Skinks

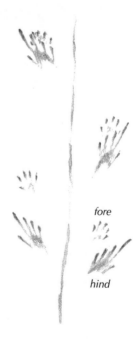

fore

hind

Straddle
to 3 in (7.5 cm)

walking

SKINKS

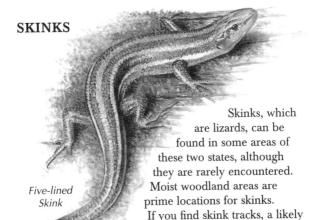

Five-lined Skink

Skinks, which are lizards, can be found in some areas of these two states, although they are rarely encountered. Moist woodland areas are prime locations for skinks.

If you find skink tracks, a likely candidate in eastern New York and most of Pennsylvania is the Five-lined Skink (*Eumeces fasciatus*), which favors moist woodlands and can grow to 8 inches (20 cm) long. Another skink that you might encounter is the Coal Skink (*E. anthracinus*); look for its tracks in mud under damp forest vegetation. It is found in the central parts of both states. In southeastern Pennsylvania you may also find tracks of the Broadhead Skink (*E. laticeps*), which grows up to 13 inches (33 cm) in length. If you are lucky enough to see a juvenile of this species, you will be rewarded by the sight of its brilliant blue tail.

Skink tracks are similar to salamander tracks, but with longer, more slender toes. Skinks, however, move very quickly when the need arises, their feet barely touching the ground as they dart for cover, so consequently their tracks can be hard to make out clearly.

Turtles

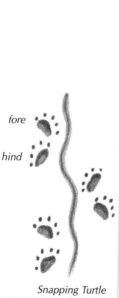

fore

hind

Straddle
4–10 in (10–25 cm)

*Snapping Turtle
walking*

fore

hind

*typical turtle
walking*

TURTLES

Painted Turtle

Turtles, those ancient inhabitants of the water world, will happily slip into the murky depths to avoid detection, but they do come out from time to time to feed or to bask in the sunshine. Look for their distinctive tracks alongside ponds, rivers and moist areas. Some turtles, such as the huge Common Snapping Turtle (*Chelydra serpentina*), prefer to stay in the water and rarely come out. Often seen basking, the Painted Turtle (*Chrysemys picta*), which can be almost 10 inches (25 cm) long, is one of the most widespread and prettiest turtles. Slightly smaller, the well-named Common Musk Turtle (*Sternotherus odoratus*) lives in shallow water and produces a foul odor. The distinctive Spotted Turtle (*Clemmys guttata*), at only 5 inches (13 cm) in length, is a common resident of beaver ponds. The Eastern Box Turtle (*Terrapene carolina*) is common in southern Pennsylvania's forested areas.

With its large shell and short legs, a turtle leaves a track that is wide relative to the length of its stride—its straddle is about half its body length. Although longer-legged turtles can raise their shells off the ground, short-legged species may let them drag, which is shown in their tracks. The tail may leave a straight dragline in the mud. On firmer surfaces, look for distinct claw marks.

Snakes

SNAKES

*Common
Garter Snake*

Many snake species inhabit the region, with much greater diversity in the warmer south. Because snakes are all long and slender, their tracks appear so similar that identification among the species is next to impossible. In fact, because a snake lacks feet and leaves a track that is just a gentle meander, it is very challenging even to establish in which direction a snake was moving.

The most widespread, frequently encountered snake is the harmless Common Garter Snake (*Thamnophis sirtalis*). Found throughout the region, often close to wet or moist areas, it can reach 4.3 feet (1.3 m) in length. Also widespread is the Redbelly Snake (*Storeria occipitomaculata*), which prefers hilly woodlands and can grow to 16 inches (40 cm) long. Inhabiting grassy meadows and fields along forest edges, the Smooth Green Snake (*Liochlorophis vernalis*) can grow to 26 inches (66 cm) long. The rattlesnake most likely to be encountered is the Timber Rattlesnake (*Crotalus horridus*), which can grow to be 6.3 feet (1.9 m) long; it frequents a variety of habitats from marshlands to dry woodlands. Also common in a variety of habitats is the beautiful Milk Snake (*Lampropeltis triangulum*).

TRACK PATTERNS & PRINTS

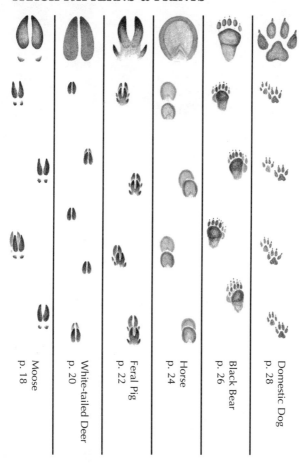

Moose
p. 18

White-tailed Deer
p. 20

Feral Pig
p. 22

Horse
p. 24

Black Bear
p. 26

Domestic Dog
p. 28

Coyote
p. 30

Red Fox
p. 32

Gray Fox
p. 34

Lynx
p. 36

Bobcat
p. 38

Domestic Cat
p. 40

TRACK PATTERNS & PRINTS

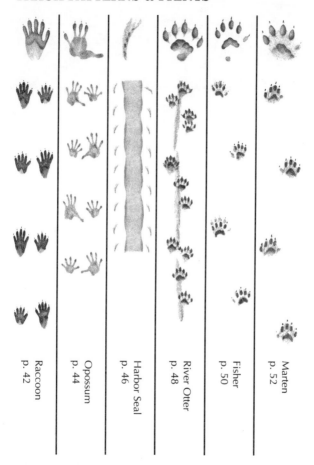

Raccoon
p. 42

Opossum
p. 44

Harbor Seal
p. 46

River Otter
p. 48

Fisher
p. 50

Marten
p. 52

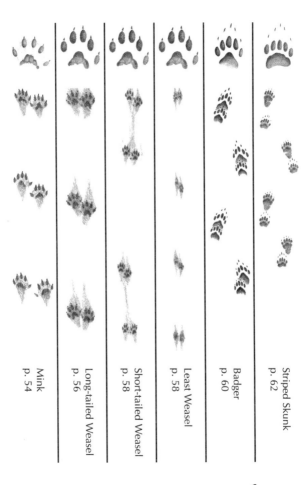

Mink
p. 54

Long-tailed Weasel
p. 56

Short-tailed Weasel
p. 58

Least Weasel
p. 58

Badger
p. 60

Striped Skunk
p. 62

143

TRACK PATTERNS & PRINTS

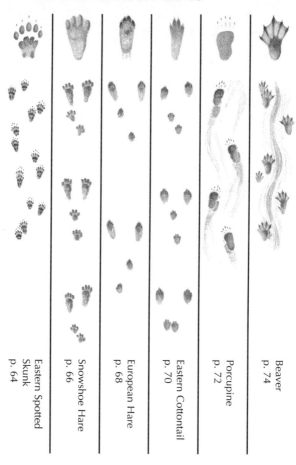

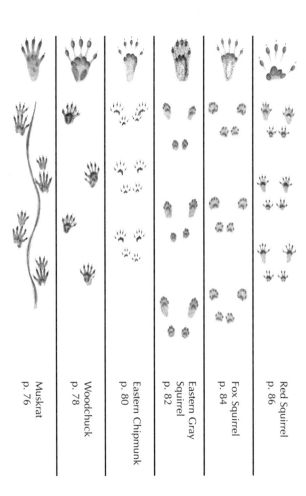

Muskrat
p. 76

Woodchuck
p. 78

Eastern Chipmunk
p. 80

Eastern Gray
Squirrel
p. 82

Fox Squirrel
p. 84

Red Squirrel
p. 86

TRACK PATTERNS & PRINTS

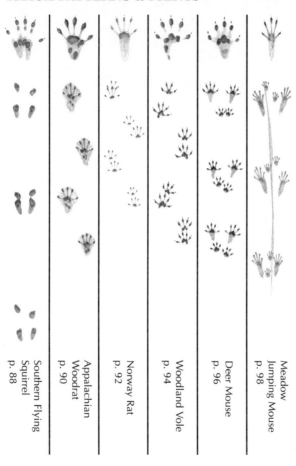

Southern Flying
Squirrel
p. 88

Appalachian
Woodrat
p. 90

Norway Rat
p. 92

Woodland Vole
p. 94

Deer Mouse
p. 96

Meadow
Jumping Mouse
p. 98

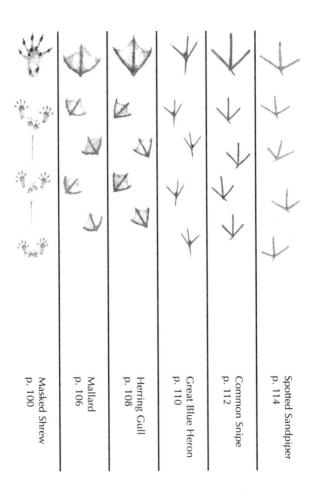

TRACK PATTERNS & PRINTS

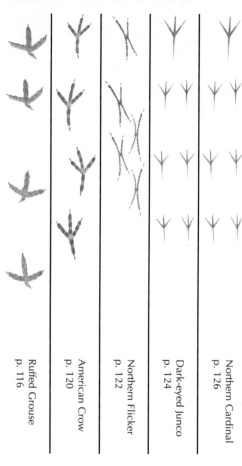

Ruffed Grouse
p. 116

American Crow
p. 120

Northern Flicker
p. 122

Dark-eyed Junco
p. 124

Northern Cardinal
p. 126

148

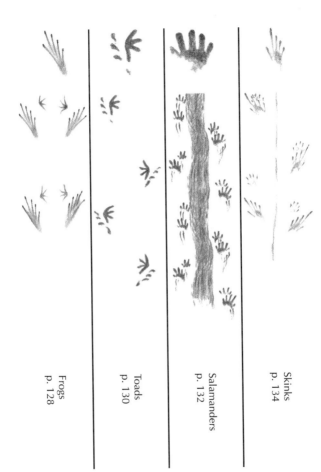

Frogs
p. 128

Toads
p. 130

Salamanders
p. 132

Skinks
p. 134

149

TRACK PATTERNS & PRINTS

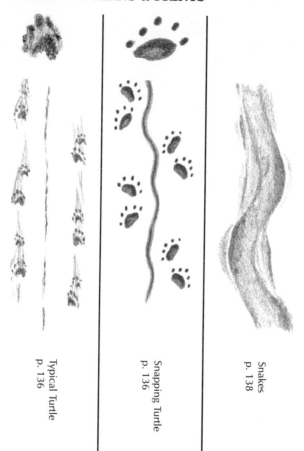

Typical Turtle
p. 136

Snapping Turtle
p. 136

Snakes
p. 138

HOOFED PRINTS

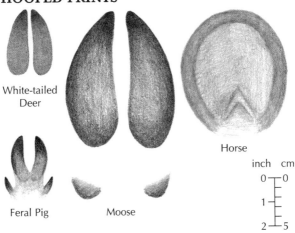

White-tailed
Deer

Horse

Feral Pig

Moose

inch   cm

0 — 0

1

2 — 5

HIND PRINTS

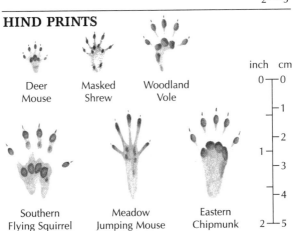

Deer
Mouse

Masked
Shrew

Woodland
Vole

inch   cm

0 — 0

1

2

1

3

4

2 — 5

Southern
Flying Squirrel

Meadow
Jumping Mouse

Eastern
Chipmunk

HIND PRINTS

Norway
Rat

Appalachian
Woodrat

Red
Squirrel

Woodchuck

Muskrat

Fox
Squirrel

Eastern Gray
Squirrel

Eastern
Cottontail

inch cm
0 ┬ 0
1 ┤
2 ┴ 5

Raccoon

Opossum

Porcupine

HIND PRINTS

inch cm
0 ┬ 0
2 ┤
4 ┴ 10

Snowshoe
Hare

European
Hare

Beaver

Black Bear

FORE PRINTS

Least Weasel

Short-tailed Weasel

Long-tailed Weasel

Domestic Cat

Gray Fox

Red Fox

Eastern Spotted Skunk

Mink

Marten

Bobcat

Coyote

Striped Skunk

Badger

River Otter

Domestic Dog

Fisher

Lynx

inch cm

0 — 0

1 —

2 — 5

BIBLIOGRAPHY

Behler, J.L., and F.W. King. 1979. *Field Guide to North American Reptiles and Amphibians.* National Audubon Society. New York: Alfred A. Knopf.

Brown, R., J. Ferguson, M. Lawrence and D. Lees. 1987. *Tracks and Signs of the Birds of Britain and Europe: An Identification Guide.* London: Christopher Helm.

Burt, W.H. 1976. *A Field Guide to the Mammals.* Boston: Houghton Mifflin Company.

Farrand, J., Jr. 1995. *Familiar Animal Tracks of North America.* National Audubon Society Pocket Guide. New York: Alfred A. Knopf.

Forrest, L.R. 1988. *Field Guide to Tracking Animals in Snow.* Harrisburg: Stackpole Books.

Halfpenny, J. 1986. *A Field Guide to Mammal Tracking in North America.* Boulder: Johnson Publishing Company.

Headstrom, R. 1971. *Identifying Animal Tracks.* Toronto: General Publishing Company.

Murie, O.J. 1974. *A Field Guide to Animal Tracks.* The Peterson Field Guide Series. Boston: Houghton Mifflin Company.

Rezendes, P. 1992. *Tracking and the Art of Seeing: How to Read Animal Tracks and Signs.* Vermont: Camden House Publishing.

Stall, C. 1989. *Animal Tracks of the Rocky Mountains.* Seattle: The Mountaineers.

Stokes, D., and L. Stokes. 1986. *A Guide to Animal Tracking and Behaviour.* Toronto: Little, Brown and Company.

Wassink, J.L. 1993. *Mammals of the Central Rockies.* Missoula: Mountain Press Publishing Company.

Whitaker, J.O., Jr. 1996. *National Audubon Society Field Guide to North American Mammals.* New York: Alfred A. Knopf.

INDEX

Page numbers in **boldface** type refer to the primary (illustrated) treatments of animal species and their tracks.

ABOUT THE AUTHORS

Tamara Eder, equipped from the age of six with a canoe, a dip net and a note pad, grew up with a fascination for nature and the diversity of life. She has a degree in environmental conservation sciences and has photographed and written about the biodiversity in Bermuda, the Galapagos Islands, the Amazon Basin, China, Tibet, Vietnam, Thailand and Malaysia.

Ian Sheldon, an accomplished artist, naturalist and educator, has lived in South Africa, Singapore, Britain and Canada. Caught collecting caterpillars at the age of three, he has been exposed to the beauty and diversity of nature ever since. He was educated at Cambridge University and the University of Alberta. When he is not in the tropics working on conservation projects or immersing himself in our beautiful wilderness, he is sharing his love for nature. Ian enjoys communicating this passion through the visual arts and the written word.